Methane Hydrates in Quaternary Climate Change

The Clathrate Gun Hypothesis

James P. Kennett
Kevin G. Cannariato
Ingrid L. Hendy
Richard J. Behl

 American Geophysical Union
Washington, DC

Library of Congress Cataloging-in-Publication Data
Methane hydrates in Quaternary climate change: the clathrate gun hypothesis / James P. Kennett...[et al.].
    p.cm.
  Includes bibliographical references.
  ISBN 0-87590-296-0
  1. Paleoclimatology--Quaternary. 2. Climatic changes. 3. Methane. I. Title: Clathrate gun hypotheses. II. Kennett, James P.
QC884.M44 2002
551.6--dc21   2002035623

ISBN 0-87590-296-0

Cover photograph: Methane hydrate mound (~1.5 m diameter) on ocean floor (540 m) at Bush Hill, northern Gulf of Mexico, with free gas venting into the water column. Mound is surrounded by chemosynthetic tube worms and covered by *Beggiatoa* mats. Courtesy of Ian R. MacDonald, Geochemical and Environmental Research Group,Texas A&M University (see Sassen, R. et al., 1998, *Geology* 26, p 851-854).
Back cover graphics: Ribbons are from a sonar reflection profile (3.5 kHz) through active hydrocarbon gas seeps at Coal Oil Point, Santa Barbara, California. Dark blue curtains are gas bubble plumes rising from the ocean floor. Water depth ~60 meters. Courtesy of Bruce Luyendyk, Hydrocarbon Seeps Study Group, University of California Santa Barbara.

Contents

Preface

Like most Earth scientists, we are intrigued and amazed by recent discoveries from ice-core and marine sediments that global climate and the ocean-atmosphere system can abruptly switch from glacial to near-interglacial temperatures within decades. Remarkably, this happened many times during and at the end of the last glacial episode, causing enormous disruptions in the global biosphere. Such discoveries are double-edged, however. Along with the excitement they prompt comes a grand challenge: their explanation. We, and others, have wondered what factors can possibly drive the climate so far and so fast. Where does the energy come from? Understanding such phenomena becomes paramount in a world with increasing concern about the potential role of humans on global climate change.

This book addresses such issues head on, if by an unconventional tack. We suggest that each of the numerous abrupt warnings during the last ice age was accompanied by, and in part driven by, massive releases of methane into the ocean and atmosphere system by dissociation of "frozen" methane hydrates (clathrates) stored in marine sediments on the continental margins. A general consensus now exists that methane hydrates represent an enormous reservoir of carbon in the form of frozen methane that both stores and has the capability of suddenly releasing free methane into the ocean and atmosphere when environmental conditions are suitable. We also suggest that repeated emissions of methane (a potent greenhouse gas) promulgated other rapid feedbacks that assisted in driving the Earth's climate abruptly from glacial to interglacial temperatures. We refer to this hypothesis as the Clathrate Gun Hypothesis.

The Clathrate Gun Hypothesis invokes processes that have been relatively inactive and unobserved by human beings during our cultural development and recorded history—the last 7000 years. This period—the latter part of the Holocene interglacial—was a time of high sea levels and warm and relatively stable atmospheric and ocean temperatures. Such conditions were ideal for methane hydrate stability, and consequently, major emissions from the reservoir are outside the modern human experience. Before the middle Holocene, however, our human ancestors experienced dramatic shifts in climate and sea level, and some may have even observed and experienced the effects of massive methane releases into the ocean/atmosphere in near-coastal areas.

Yet, because the methane hydrate reservoir has been largely stable during recorded human experience, its potential influence is little incorporated into general models of past climate change. A recent major synthesis of abrupt climate change relegates the potential significance of methane hydrates to the category of "exotica and surprises." We believe that if methane hydrates were a current source of large methane emissions into the atmosphere, they almost assuredly would have been long considered an integral part of the climate system, which we suggest they are. Herein, we present much evidence that the methane hydrate reservoir has vigorously interacted with the ocean and atmosphere in the recent geological past.

Key discoveries made during the last decade have led scientists to recognize the importance of three areas of Earth science discussed in this book: (1) the extreme sensitivity of the Earth's environmental system to change during the last ice age; (2) the abruptness and near-global extent of major climate and environmental change that occurred over decades; and (3) an appreciation of the extent of the methane hydrate reservoir as a major carbon reservoir. This book presents an assessment that links all three into an integrated global hypothesis. It invokes inherently unstable methane hydrates as a critical component of the climate system in providing crucially needed energy to force and accelerate abrupt climate change.

Our interest in the potential relations between climate change and methane hydrates derives from our studies of the dramatic paleoclimatic and paleoceanographic change recorded and contained in the methane-rich sediments of Santa Barbara Basin. Specifically, we began to formulate the ideas outlined in this book upon discovery in 1995 of a sequence of brief, highly negative carbon isotopic spikes exhibited in fossil planktonic and benthic foraminifers in Santa Barbara Basin sediments. These immediately reminded us of a carbon isotopic spike in foraminifera discovered by Lowell Stott and Jim Kennett associated with the terminal Paleocene thermal maximum and major deep-sea extinction 57 million years ago, which Jerry Dickens later attributed to massive dissociation of marine methane hydrates due to deep ocean warming. We still find it remarkable that discoveries of the Earth's operating style in its ancient past provide clues and critical insight into key climatic processes that may have operated in the near-recent geologic past, providing yet another example of the importance of the past as a key to the present and future.

The present book integrates and summarizes widely disparate research fields in a process of consilience; although no single proof exists for the hypothesis presented here, a wide range of evidence supports it. For this reason alone, we hope that the book interests a broad range of Earth scientists. We expect it to be especially appropriate for scientists compelled by the history, and processes, of the Earth as an integrated system but coming from different approaches, including: paleoceanography, climate dynamics, paleobotany, biogeochemical cycling, methane hydrates, and marine geology. In formulating this integrated hypothesis we recognize that different aspects are controversial. However, we hope that the present hypothesis stimulates further research on possible linkages and significance. To this end, we have outlined a number of research areas and tests that may clarify relationships and promote better understanding of the causes of late Quaternary climate change.

Acknowledgments: I began writing this manuscript while on sabbatical leave as a guest in the Department of Earth and Ocean Sciences at the University of British Columbia, Canada. I thank the Department, and especially Drs Tom Pedersen, Steven Calvert and Richard Chase for their wonderful hospitality and for helping make my sabbatical so stimulating and productive. I also thank Lionel Carter, Keith Lewis, and NIWA in Wellington, New Zealand, for their help and hospitality during initial stages of this project in early 2000. My colleagues and I also received assistance from several individuals at the University of California, Santa Barbara. In particular, we thank Karen Thompson for considerable assistance in gathering and compiling references, Jo Anne Sharpe for secretarial assistance in preparing the first draft, and Diana Kennett for assisting in so many ways. Initial stimulation for this investigation resulted from the innovations of G. Dickens and E. Nisbet. We thank numerous other colleagues for providing useful suggestions, stimulation, or constructive criticism as this research evolved. These individuals include P. Brewer, H. Behling, B. Buffett, T. Blunier, J. Chappell, J. Clark, G. Denton, P. Glaser, T. Hill, K. Hinrichs, R. Hyndman, C. Keeling, L. Keigwin, M. Kienast, K. Kvenvolden, D. Lea, I. Leifer, B. Luyendyk, M. Maslin, H. Nelson, U. Ninnemann, C. Paull, D. Piper, W. Reiners, M. Sarnthein, J. Schimel, J. Severinghaus, C. Sorlien, L. Talley, and R. Wüst. We also wish to thank L. Peterson, C. Ruppel, and an anonymous reviewer for their most valuable constructive suggestions for improvements upon reading of the entire manuscript. The sections on methanogenesis and methanotrophy were considerably improved as a result of constructive suggestions by D. L. Valentine.

Dr. Dale Krause provided much stimulation, encouragement, and valuable advice throughout while offering useful comments on the manuscript. We also appreciate the efforts of Allan Graubard, Acquisitions Editor of the American Geophysical Union books program, for constructive suggestions and encouragement throughout and for his guidance through the review process; in effect, for making this publication a reality. We also thank Terence Mulligan, our production editor, for carefully transforming our manuscript into this volume. This research was initially inspired as a result of Ocean Drilling Program (ODP) participation. The ODP is sponsored by the U.S. National Science Foundation (NSF) and participating countries under the management of Joint Oceanographic Institutions (JOI) Inc. Funding for this research was provided by the JOI/U.S. Science Support Program and NSF (Marine Geology and Geophysics) and we especially thank Dr. Bil Haq for his help and encouragement. Richard Behl wishes to make acknowledgment to the Donors of the American Chemical Society Petroleum Research Fund, for partial support of this research.

About the Authors

James Kennett, a marine geologist, has been a Professor in the Department of Geological Sciences, University of California, Santa Barbara, since 1987, and was Director of the Marine Science Institute at UCSB from 1987 to 1997. He received his PhD (1965) and DSc (1976) from Victoria University of Wellington, New Zealand.

Kennett has published extensively in stratigraphy, marine micropaleontology, marine geology and contributed towards the development of paleoceanography. He is a Fellow of the American Geophysical Union (AGU), the American Association for the Advancement of Science and the Geological Society of America as well as a Member of the U.S. National Academy of Sciences and Honorary Member of the Royal Society of New Zealand. He has had numerous associations with AGU including being Foundation Editor of Paleoceanography and election in 2000 as Emiliani Lecturer by the Ocean Sciences Section of AGU.

His contributions towards abrupt climate change during the Quaternary have continued since 1992 with the successful coring of the outstanding Santa Barbara Basin sequence by the Ocean Drilling Program. Results from this research stimulated his interest in the potential role of methane hydrates in climate change.

He and his wife Diana have lived in the U.S. since 1966 and have two grown children.

Kevin Cannariato obtained a PhD (2002) in paleoceanography in the Department of Geological Sciences, University of California, Santa Barbara. Currently, he is a Post-doctoral Fellow in the Department of Geological Sciences, University of Southern California. His primary interests are causes of climate change and affects on the biota of past environmental changes.

Ingrid Hendy obtained a PhD (2000) in paleoceanography in the Department of Geological Sciences, University of California, Santa Barbara. She was a Post-doctoral Fellow (2000-2002) at the University of British Columbia, Vancouver, Canada, and is currently an Assistant Professor in the Department of Geological Sciences, University of Michigan. She received the Doris Curtis Women in Science Award in 2001 by the Geological Society of America in recognition of her paleoclimatic research in Santa Barbara Basin. Her interests center on past ocean and climate change as studied from marine sediment records.

Richard Behl is Associate Professor in the Department of Geological Sciences, California State University, Long Beach, where he has taught since 1995. He obtained a PhD in Earth Sciences at the University of California, Santa Cruz (1992), having researched the sedimentology and diagenesis of siliceous sediments along the California margin and in the deep Pacific Ocean. From 1992 to 1995 he was a Post-doctoral Fellow in the Marine Science Institute, University of California, Santa Barbara, where he began research on Quaternary climate change as recorded in Santa Barbara Basin. His primary interests are in marine sedimentology including diagenesis, geochemistry, petroleum geology, paleoceanography and climate change.

Methane Hydrates in Quaternary Climate Change: The Clathrate Gun Hypothesis

James P. Kennett, Kevin G. Cannariato, Ingrid L. Hendy,
and Richard J. Behl

SUMMARY

The remarkable similarity of late Quaternary atmospheric methane (CH_4) and temperature variations recorded in ice cores suggests that CH_4 has played a significant role in climate change. Dissociation of marine sedimentary methane hydrates likely induced episodes of rapid climate warming at various times in the geologic past through greenhouse forcing by atmospheric CH_4. We propose that the late Quaternary (last 800 kyr) was also a time of significant instability of the methane hydrate reservoir. This sensitivity resulted from the interaction of large changes in sea level (pressure) and fluctuating character of intermediate water (temperature) impinging on upper continental slopes (~400 to 1000 m water depth); the zone of potential methane hydrate instability. After 800 ka, upper continental margin methane hydrate reservoirs became more strongly engaged in greenhouse amplification of global climate change. This occurred when colder glacial upper intermediate waters expanded methane hydrate accumulation into shallower waters at the same time that intermediate-water temperatures became more variable.

According to the *Clathrate Gun Hypothesis*, episodic CH_4 release resulting from the dissociation of the sedimentary methane hydrate reservoir contributed significantly to the distinctive behavior of late Quaternary climate change on orbital (Milankovitch) and millennial time scales. Repeated changes between sources and fluxes of upper intermediate waters caused fluctuations in temperature that alternately accumulated and episodically dissociated hydrates on upper

Methane Hydrates in Quaternary Climate Change
© 2003 by the American Geophysical Union
10.1029/054SP01

continental slopes. The Clathrate Gun metaphor refers to the sequential and repeated steps of loading the methane hydrate reservoir, triggering hydrate dissociation by changes in intermediate-water temperature, and energetically discharging CH_4 into the ocean/atmosphere to amplify late Quaternary climate warming events that were strongly reinforced by other greenhouse gases, especially water vapor.

According to this hypothesis, methane hydrates stabilized and accumulated during late Quaternary cool intervals when cold intermediate waters bathed upper continental margins. Coldest intervals occurred when reinforcement of orbital insolation cycles led to largest ice sheets and the greatest accumulation of methane hydrates. Changes in thermohaline circulation that caused warming of upper intermediate waters resulted in methane hydrate instability and catastrophic release of CH_4 into the ocean/atmosphere system associated with sediment disruption on upper continental slopes. Episodes of major CH_4 release from hydrates appear to have been recorded in late Quaternary sediment sequences on different continental margins by very negative $\delta^{13}C$ excursions in benthic and planktonic foraminifera. It is suggested that these releases, in association with other climatic feedback mechanisms, triggered rapid warmings of different magnitude and contributed to the distinctive sawtooth pattern of late Quaternary climate change exhibited over a wide range of time scales including the 100-kyr cycle, Bond Cycles, and individual interstadial events (Dansgaard/Oeschger interstadials).

Much geological evidence indicates that the abrupt CH_4 increases at glacial and stadial terminations were more likely to have been produced by methane hydrates than by continental wetlands. Inundated flood plains and organic carbon buildup, necessary for major CH_4 production, had not significantly developed during glacial lowstands of sea level, because incised rivers resulted in vigorous freshwater discharge from the continents. These wetland systems were insufficiently established during rapid warmings to have produced the dramatic atmospheric CH_4 increases that occurred within decades. The large, complex modern wetland ecosystems (peatlands, tropical floodplains, coastal wetlands) did not begin to develop substantially until the middle Holocene because of low sea level, low water tables and dryness over large areas of the Earth. Extensive modern high latitude wetlands were not established until after the ice sheets and glaciers of the high northern latitudes melted, well after the glacial terminations. Major wetlands in non-glaciated high latitude areas also did not develop until within the Holocene. Furthermore, much paleobotanical and other data indicate that tropical wetlands only developed extensively after the middle Holocene, following sea-level rise to modern levels, infilling of incised river channels and formation of flood plains. In the absence of established CH_4-producing wetlands, the rapid increases in atmospheric CH_4 recorded in ice cores are more consistent with a methane hydrate source.

Abrupt atmospheric CH_4 increases at glacial and stadial terminations suggest that the greatest CH_4 expulsions from the methane hydrate reservoir occurred at or near the beginning of warm episodes. Initial abrupt increases were followed by slower declines as the methane hydrate reservoir equilibrated with higher sea levels and warmer temperatures, as predicted by the hypothesis. Dynamic release of CH_4 is supported by conspicuous CH_4 overshoots during initial rapid CH_4 rises at the onset of several warm intervals. In contrast, continental wetland sources of CH_4 would exhibit opposite behavior from that observed in the ice cores; a general increase in CH_4 during warm, wet intervals as major wetland ecosystems and necessary organic carbon accumulations built up, such as what occurred during the middle to late Holocene.

Late Quaternary climate oscillations led to successive intervals of methane hydrate stability and accumulation during coolings, followed by hydrate dissociation and CH_4 release during warmings. These changes occurred because of frequent, rapid temperature oscillations in upper intermediate waters over broad areas of the upper continental margins in the depth zone of potential hydrate instability. We present evidence suggesting that these temperature oscillations resulted from changes in the source of intermediate waters between low and high latitudes. Cool episodes were marked by increased equatorward expansion of cooler intermediate water from northern (North Pacific and North Atlantic Intermediate Waters) and southern latitudes (Antarctic Intermediate Water).

Much evidence exists for extensive instability of upper continental slope sediment during the late Quaternary related to methane hydrate dissociation. This instability is indicated by widespread development of slumps, debris flows and pockmarks, with associated transport of terrigenous sediment to ocean basins forming turbidity, nepheloid and other sedimentary deposits. Slope instability during the latest Quaternary appears not to have peaked during lowest sea levels of the last glacial maximum, but when sea level was rising most rapidly during the following deglacial episode, at the time that the methane hydrate reservoir is inferred to have been most unstable. Methane hydrates at intermediate water depths were especially destabilized by a widespread switch to warmer bottom waters after prolonged major cooling and methane hydrate build-up at a time when sea level (pressure) was lowest. The hypothesis predicts the large CH_4 emissions at glacial terminations that are documented in ice cores.

The pervasive sawtooth pattern of late Quaternary climate change on orbital to millennial time scales suggests a climate system marked by feedbacks related to the stability or instability of the methane hydrate reservoir. We suggest that this pattern resulted from repeated cycles of methane hydrate accumulation within shallower depth ranges on the continental margins that subsequently became sensitive to fluctuation of intermediate-water temperature. The amplitudes of these cycles were modulated by Milankovitch cyclicity that set the timing and magnitude of greatest cooling (optimal conditions for hydrate accumulation). Subsequent warming of even minor magnitude triggered a sequence of positive feedbacks. Release of methane hydrate-derived CH_4 into the atmosphere initiated a cascade of feedbacks, especially on short time scales: greenhouse warming by atmospheric CH_4 and water vapor, warming and expansion of intermediate waters, additional methane hydrate dissociation over broader depths and regions, and so on, all at a time when sea level and confining pressure were lowest. Gradual coolings in the sawtooth followed dissociation of most readily accessible methane hydrates, depletion of atmospheric methane, and weakening of associated global warming feedback mechanisms. Re-expansion of methane hydrates into shallow sediment and water depths, the zone of potential methane hydrate instability, only followed sufficient cooling of upper intermediate waters.

The *Clathrate Gun Hypothesis* contributes to the explanation of several phenomena of the late Quaternary climate record. These include: tightly-linked climate changes over broad areas in both hemispheres; abrupt warmings and coolings; the sawtooth pattern of climate behavior marked by warmest intervals immediately following coldest intervals; the similar behavior of the ice core climate and CH_4 records; the dominance of the 100-kyr cycle, and its beginnings at ~800 ka; the so-called Stage 11 problem, and the increased amplitude of climate change during the late Quaternary.

INTRODUCTION

Understanding the causes of late Quaternary climate behavior represents one of the major challenges in earth sciences. High-resolution studies of climate change using ice cores [e.g., Dansgaard et al., 1993; Grootes and Stuiver, 1997], and marine [e.g., Bond et al., 1993; Schulz et al., 1998; Hughen et al., 1998; Sachs and Lehman, 1999; Hendy and Kennett, 1999, 2000; Kennett et al., 2000a; Peterson et al., 2000] and terrestrial sediments [e.g., Allen et al., 1999] have revealed rapid, large millennial-scale climate oscillations during the last glacial episode (see Voelker et al. [2002] for detailed list) (Figure 1). These studies describe an almost bistable oscillatory behavior of the climate system, with switches between distinctly different states occurring in decades or less [Alley and Clark, 1999]. Even more remarkable is the amplitude of temperature changes associated with these high-frequency climate shifts, at times, reaching glacial-interglacial magnitudes (Figure 1) [Dansgaard et al., 1993; Alley and Clark, 1999]. Each of the warm events (interglacials and interstadials) during the late Quaternary required a trigger to initiate the change, strong positive feedbacks for reinforcement, and a process that maintained the new, nearly stable state [Nisbet, 1990]. Major positive feedbacks were necessary to produce such large, rapid shifts in global climate, but the nature of these feedbacks has remained enigmatic. A recent National Research Council [2002] review of the character, causes, and consequences of abrupt climate change concluded that this climate behavior still remains to be explained. Furthermore, climate models have typically underestimated the magnitude, speed, and extent of these abrupt changes.

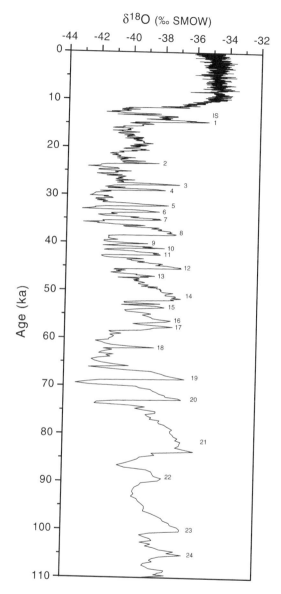

Figure 1. History of millennial-scale, rapid climate change during the last 110 kyr as recorded by the GISP2 $\delta^{18}O_{ice}$ time series (reflecting air temperature and changes in source or seasonality of precipitation) recovered from the Greenland ice sheet [Grootes, et al., 1993; Stuiver et al., 1995; chronology after Meese et al., 1994 and Sowers et al., 1993]. Climate instability was pronounced during the last glacial episode. The record exhibits a series of warm phases, the Dansgaard-Oeschger (D/O) interstadials (numbered), that punctuate the otherwise cold conditions of the last glacial episode.

Hypotheses proposed to account for the remarkable late Quaternary climate behavior are numerous and diverse, but usually invoke forcing by greenhouse gases (e.g., CO_2, CH_4, N_2O, H_2O) [Petit et al., 1999] and/or rapid shifts in ocean thermohaline circulation [Broecker, 1997a; Keeling and Stephens, 2001]. Any mechanism proposed to explain late Quaternary climate behavior must explain a diversity of documented changes, including: the character, magnitude, and speed of climate change, synchronism and diachronism between different regions and components of the earth system, changes in atmospheric greenhouse gas composition, deep and intermediate-water circulation, ecosystem changes influencing global biogeochemical cycling, and the leads and lags between the different components. So far, no proposed hypothesis has successfully explained the wide range of climate changes that occurred on orbital, millennial, and decadal time scales during the late Quaternary [Paillard, 2001].

One mechanism for late Quaternary climate change that has received little attention by the paleoclimate community is the potential role of greenhouse forcing by atmospheric CH_4 (methane) emissions and associated feedbacks resulting from instability of the marine sedimentary methane hydrate reservoir. At present, methane hydrates form a vast, apparently stable CH_4 reservoir, thus playing only a minor role in the modern CH_4 cycle [Cicerone and Oremland, 1988; Judd, 2000; Sassen et al., 2001a, b]. However, this may not have been the case throughout earth history, as increased evidence suggests that important paleoclimate events were associated with atmospheric CH_4 emissions from methane hydrates [e.g., Dickens et al., 1995]. The purpose of this contribution is to present the hypothesis that instability of the methane hydrate reservoir and resulting changes in atmospheric CH_4 concentrations played a critical role in late Quaternary climate change. Methane is the most abundant atmospheric organic compound. Atmospheric CH_4 has been shown to be an exceedingly powerful greenhouse gas [Hanson and Hanson, 1996]—62 times the Greenhouse Warming Potential of CO_2 [Ehhalt et al., 2001] on the short time scales (decades) of late Quaternary abrupt warmings [Alley and Clark, 1999]. We will demonstrate the consistency of this hypothesis in explaining a broad range of geologic data, and identify possible inconsistencies. We suggest that the synthesis presented in this contribution makes a compelling case. In essence, the hypothesis we present for further testing has resulted from the recognition of a wide range of independent and disparate results that are suggestive of a methane hydrate explanation—a strategy termed consilience [Gould, 1989; Wilson, 1998].

Valuable general reviews about methane hydrates have been presented by Kvenvolden [1988b; 1993], MacDonald [1990a; 1992], Haq [1998b], and Buffett [2000], and by many authors in Max [2000]. Concern about the possible influence of hydrate-derived CH_4 on global warming led MacDonald [1982] to consider the potential role of methane hydrate instability in climate change. Bell [1982] and Revelle [1983] followed with the first formal analyses of the problem. Nisbet [1990, 1992] introduced the hypothesis that hydrates played a major role in Quaternary climate change. He suggested that CH_4 released from high latitude hydrates at the end of the last glacial episode forced a rapid rise in global temperatures. Under this hypothesis, initial deglacial warming resulted from CH_4 emissions from melting Arctic permafrost, in turn leading to a cascade of methane hydrate release from ocean sediments. This rapidly provided new energy to the climate system, abruptly driving it to a warmer state. Several other workers have implicated methane hydrates in past climate change [Kvenvolden, 1993; MacDonald, 1990a; Paull et al., 1994; Haq, 1998a, b, 2000; Dickens et al., 1995, 1997; Kennett et al., 1996, 2000a, b]. Early support for methane hydrate instability as a potential mechanism for climate change came from the realization that decreased pressurization of the methane hydrate reservoir by sea-level lowering can destabilize hydrates [Nisbet, 1990; Paull et al., 1995; Haq, 1993]. More recent work suggests that bottom-water temperature changes may play a crucial role in methane hydrate destabilization [Dickens et al., 1995, 1997; Dickens and Quinby-Hunt, 1997; Kennett et al., 2000a; Haq, 2000].

Despite the potential for large, episodic releases of CH_4 by hydrate dissociation, intense ongoing investigations of late Quaternary climate behavior have not included the methane hydrate reservoir as an integral part of the global climate system. This is surprising given the immense amount of CH_4 estimated to be stored in the modern reservoir [MacDonald, 1990a; Kvenvolden, 1988a, b; 1993]. Paull et al. [1995] stated that "if these natural gas hydrate deposits are dynamic reservoirs, the potential to affect the Earth's climate by releasing CH_4 to the atmosphere has to be considered." The methane hydrate reservoir is generally not considered an important component in late Quaternary climate change for several reasons: the reservoir is remote and poorly studied; little was known about methane hydrates until recently; and modern hydrates appear stable and show little evidence during the Holocene of prior instability earlier in the Quaternary. Finally, the upper continental margin, containing most of the reservoir, was considered a relatively stable oceanic environment, uninvolved with Quaternary climate behavior, especially on millennial time scales. For these and other reasons, methane hydrates have been considered "a last resort" hypothesis for climate change [K. Kvenvolden, personal communication]. Nevertheless, Buffett [2000] has suggested that methane hydrates may be the "dark horse" of global climate change.

Interest in the potential role of methane hydrates in late Quaternary climate change has recently increased for several reasons:

- Recognition of the potential instability of methane hydrates on continental margins because of sea-level and bottom-water temperature changes [Paull et al., 1995; Lerche and Bagirov, 1998; Kennett et al., 1996, 2000a, b].
- Widespread agreement that catastrophic CH_4 release from dissociating hydrates caused a distinct episode of global warming at the end of the

Paleocene [e.g., Dickens et al., 1995, 1997; Bains et al., 1999; Katz et al., 1999].

- Discovery of large, negative $\delta^{13}C$ excursions in late Quaternary marine sediments from the southern California margin [Kennett et al., 2000a], Gulf of California [Keigwin, 2002], East Greenland [Smith et al., 2001], and Amazon Fan [Maslin et al., 1997] attributed to CH_4 release from methane hydrates associated with millennial-scale climate change and deglaciation.

- Recognition of the close association between CH_4 and climate recorded in ice cores, both at orbital [Petit et al., 1999] and millennial time scales [Brook et al., 2000]. The inferred changes in atmospheric CH_4 concentrations have generally been linked to wetland development; potential effects of methane hydrates have been largely ignored.

- The growing recognition of the powerful Greenhouse Warming Potential of CH_4 as reinforced by reactions and feedbacks with other atmospheric gases [Ehhalt et al., 2001].

- Discovery of large, seasonal fluxes of CH_4 released from beneath sea ice in the Sea of Okhotsk [Suess et al., 1999], a region with an ocean-floor gas source [Valyashko and Demina, 1987]. Methane derived from hydrates is transported to surface waters in gas plumes [Zonenshain et al., 1987]. Kvenvolden et al. [1992] also discovered CH_4 trapped beneath sea ice off Alaska. These observations led Suess et al. [1999] to conclude that methane hydrates may be a significant source of atmospheric CH_4. Loehle [1993] considered CH_4 to be an important, but cyclic component of the global carbon cycle during the late Quaternary.

This contribution considers the possible role of marine sedimentary methane hydrates in late Quaternary climate change. The manuscript consists of twelve main sections, presented in a logical progression of ideas. Chapter 1 describes late Quaternary climate behavior, much of which has not been adequately explained with existing hypotheses. Chapter 2 summarizes behavior of CH_4 during the late Quaternary as recorded in ice cores. Changes in atmospheric CH_4 concentrations are almost exclusively interpreted by other workers to have resulted from changes in continental wetland CH_4 emissions (the *Wetland Methane Hypothesis*). If most methane was produced in wetland systems throughout the Quaternary, there should be strong geologic evidence indicating the presence of wetlands of sufficient extent to produce the major and rapid rises in CH_4 recorded in ice cores. Chapter 3 describes the processes and environments of biogenic methane formation and oxidation. Chapter 4 represents the only comprehensive global synthesis of geologic data yet assembled for late Quaternary wetland development. A synthesis of global scale is necessary to examine all potential continental CH_4 sources through the late Quaternary. Based on this synthesis, we reject the *Wetland Methane Hypothesis* proposed to explain the rapid atmospheric CH_4 increases recorded in ice cores. Wetlands of sufficient magnitude did not exist at the times of abrupt CH_4 rise. Chapter 5 suggests instead that the rapid CH_4 increases during the late Quaternary resulted from destabilization of the methane hydrate reservoir, the only other possible CH_4 source of significant magnitude. Chapter 6 summarizes major elements of the *Clathrate Gun Hypothesis* which invokes alternating episodes of methane hydrate accumulation and instability, with resulting abrupt atmospheric CH_4 increases as an integral part of climate change during the late Quaternary. Chapter 7 summarizes recent progress towards understanding the role of methane hydrate reservoir instability in pre-Quaternary climate change. Episodes involving the effects of inferred CH_4 releases on climate change range broadly from the early Cenozoic to the Neoproterozoic. These episodes have provided valuable insight about the nature and effects of methane hydrate instability on global climate and the biosphere. Chapter 8 discusses possible causes of methane hydrate instability during the late Quaternary. We examine the respective roles of sea-level and bottom-water temperature changes as destabilizing agents. We summarize available evidence on upper intermediate-water history during the latest Quaternary and conclude that associated temperature change of these waters played a critical role in late Quaternary methane hydrate instability, in conjunction with sea-level change. Chapter 9 summarizes a significant body of geologic data that supports the hypothesis that the methane hydrate reservoir has indeed undergone significant instability during the late Quaternary. Furthermore, the inferred history of this instability, although based on more limited data, appears consistent with the *Clathrate Gun Hypothesis*. Chapter 10 discusses the potential role of CH_4 in climate change and the importance of other climate feedback mechanisms that likely reinforce the effects of atmospheric CH_4 increases. Chapter 11 presents evidence for the role of methane hydrates in Quaternary climate behavior and suggests that a number of enigmatic characteristics of climate change may be partially explained by the *Clathrate Gun Hypothesis*. Finally, Chapter 12 suggests a number of tests of the *Clathrate Gun Hypothesis* and other possible future investigations of questions that have emerged as a result of this synthesis.

1

Late Quaternary Climate Patterns

Polar ice cores and marine sediments document a remarkable series of climate cycles during the late Quaternary on orbital (Milankovitch), millennial and decadal time scales (Figures 1 & 2) [Lorius et al., 1990; Dansgaard et al., 1993; Chappellaz et al., 1990; Brook et al., 1996; Raynaud et al., 2000; Imbrie et al., 1992, 1993; Hendy and Kennett, 1999, 2000; Hendy et al., 2002; Labeyrie, 2000; Sarnthein et al., 2000; Y. Wang et al., 2001; Voelker et al., 2002]. The onset of interglacial and interstadial episodes are marked by dramatic warmings that occurred in just a few decades [Severinghaus et al., 1998; Hughen et al., 1996] and involved change in a coupled earth system including the atmosphere, ocean, ice, land surface, and biosphere [Paillard, 2001]. Each of these rapid warmings was closely linked to a rapid increase in atmospheric CH_4 (Figure 2) [Petit et al., 1999; Severinghaus et al., 1998; Severinghaus and Brook, 1999]. However, the cause of these rapid warmings and many other aspects of late Quaternary climate behavior have largely remained a mystery.

On relatively long time scales of tens to hundreds of thousands of years, changes in the distribution of solar radiation associated with the Earth's orbital oscillations have provided external forcing to the climate system [Imbrie et al., 1992; 1993]. However, these climate cycles are associated with little net change in the incoming solar radiation, so large feedbacks must have been involved [Alley and Clark, 1999; Paillard, 2001]. Insolation forcing appears to act as a trigger for fundamental reorganization of the non-linear system, but does not control the magnitude of these changes [Denton et al., 1999]. Furthermore, although insolation changes are out of phase between the hemispheres, climate oscillations appear synchronous on orbital time scales over broad regions in both hemispheres. In spite of their obvious importance, the processes that amplify these astronomical variations into climate changes of considerable speed and magnitude over much of the globe have remained elusive. For example, it is difficult to account for the dominant 100-kyr cycle that marked the late Quaternary (Figure 2) with eccentricity variations

Methane Hydrates in Quaternary Climate Change
© 2003 by the American Geophysical Union
10.1029/054SP02

because the insolation changes are too small to produce the corresponding climate changes by direct forcing [Imbrie et al., 1993; Shackleton, 2000].

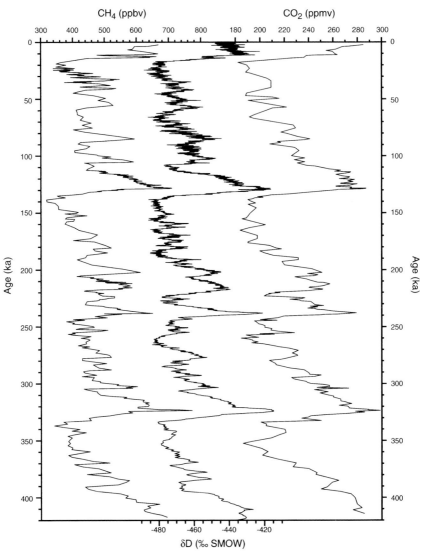

Figure 2. Glacial-interglacial variations of CH_4 (left), δD (middle), and CO_2 (right) for the last 4 glacial cycles (423 kyr) recovered from the Vostok ice core, Antarctica (depth 3,310 m in the East Antarctic ice sheet) [Petit et al., 1999]. δD is largely a measure of atmospheric temperature. Mean sample resolution of CO_2 over most of the record is 1500 yr. The resolution of the chronology (Model GT4) is better than ±10 kyr for most of the record and better than ±5 kyr for the last 110 kyr [Petit et al., 1999].

Glacial Terminations

Glacial terminations represent episodes of greatest warming and associated rapid deglaciation following each glacial episode (Figure 2). Deglaciations occurred at, or close to, maximum ice volume and lowest sea level during the glacial cycle [Shackleton, 1987]. Rapid melting of the cryosphere occurred in Northern Hemisphere ice sheets, alpine regions, and Antarctica. Terminations followed extended intervals of large ice volume and low sea level [Clark et al., 2002]. Rapid melting at the last glacial termination was caused by a remarkable rise in air temperature (~15°C over Greenland) [Dansgaard et al., 1993; Alley and Clark, 1999; Cuffey et al., 1995]. Similar dramatic warming occurred at low latitudes including an 8°C sea-surface temperature (SST) increase on the California margin [Kennett and Ingram, 1995a; Hendy and Kennett, 2000; Hendy et al., 2002], 4° to 6°C in the tropics [Stute et al., 1995], and >1°C in the South China Sea [Kienast et al., 2001]. In the high latitude Northeast Pacific, SST changes during the last termination were strongly coupled with those in the North Atlantic; their apparent synchroneity suggesting an atmospheric transmission of the climate signal [Kienast and Mckay, 2001].

During the last glacial termination (Termination 1), much of the temperature rise occurred in two steps (Plate 1), at the beginning of the Bølling-Ållerød (B-Å) warming (Termination 1A, at ~14.7 ka), and at the beginning of the Holocene (Termination 1B, at ~11.5 ka). These two warm episodes are separated by the last stadial cooling, known as the Younger Dryas [Dansgaard et al., 1993; Denton and Karlén, 1973]. High-resolution records from Greenland ice cores [Severinghaus et al., 1998; Severinghaus and Brook, 1999] have demonstrated that both temperature increases occurred remarkably rapidly (Figure 1). Termination 1A (9°C increase over Greenland) took place in only a few decades and perhaps as rapidly as 7 to 14 yr, while Termination 1B (5 to 10°C increase) occurred in less than a decade, and perhaps as little as 3 yr [Alley et al., 1993, 1995; Mayewski et al., 1997; Severinghaus et al., 1998; Severinghaus and Brook, 1999]. Thus, most of the glacial-interglacial warming took place exceptionally rapidly when the cryosphere was close to maximum extent and volume.

Nevertheless, the last glacial maximum (LGM) (21.5 to 18.3 ka) was followed by an initial warming phase of slower melting and Northern Hemisphere ice sheet retreat from ~18 ka to 14.7 ka [Bard et al., 1987; Jones and Keigwin, 1988; Denton et al., 1999; Lagerklint and Wright, 1999, and references therein]. SST rise also clearly preceded major ice volume effects on $\delta^{18}O$ in the equatorial Pacific [Lea et al., 2000], tropical Atlantic [Rühlemann et al., 1999; Nürnberg et al., 2000] and Arabian Sea [Bard et al., 1997]. Retreat of the Fennoscandian ice sheet [Denton et al., 1999] and associated deglaciation during this warming preceding the Bølling caused a 20 m sea-level rise prior to the major melt-water pulse and rapid sea-level rise that marked the beginning of the B-Å (Termination 1A) [Fairbanks, 1989; Sowers and Bender, 1995; Bender et al., 1999; Hanebuth et al.,

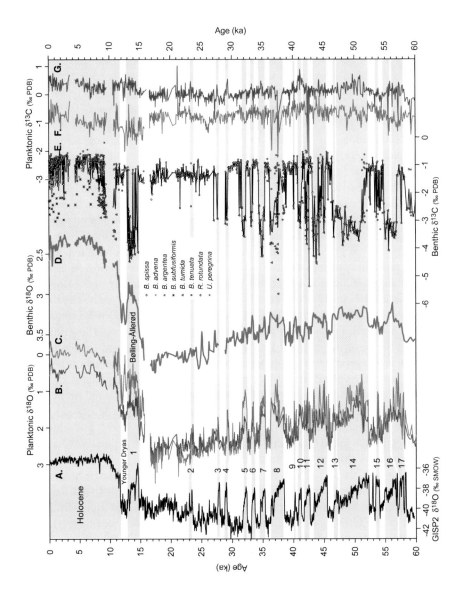

2000]. This preliminary, slow phase of melting coincided with an early interval of Antarctic warming [Bender et al., 1999; Jouzel et al., 2001; Monnin et al., 2001]. Major climate feedbacks did not reinforce this event as they did at the B-Å. However, the beginning of this episode appears to have been associated with warming of intermediate waters (<~1000 m) on the southern California margin [Kennett et al., 2000a; Hendy and Kennett, submitted; Kennett et al., in prep.] that clearly preceded the rapid warming. It is possible that this pre-Bølling warming represents a critical and necessary prelude to the massive, rapid warming at the B-Å [Denton et al., 1999], which clearly involved major positive feedbacks.

Following the terminations, peak interstadial and interglacial warmth occurred at or near the beginning of the warm episodes (Figures 1 & 2). Maximum Bølling warmth was short-lived between 12.5 and 12.1 ^{14}C kyr B.P., and was followed by a cooling trend that persisted during much of the late B-Å [Denton et al., 1999]. Likewise, the early Holocene was the warmest, forming the so-called Holocene climatic optimum [Alley et al., 1997; Haug et al., 2001].

The climate history outlined above indicates that the late Quaternary had a tendency for intervals of cooling and ice accumulation to be followed by episodes of rapid, maximum warming. Full interglacial episodes developed rapidly at the beginning of each 100-kyr cycle (Figure 2). The remainder of each cycle is marked by climate oscillations superimposed on a general cooling trend terminating in a glacial maximum. The major warming at the beginning of each cycle is very rapid while cooling toward full glacial conditions is prolonged (Figure 2). This results in asymmetrical glacial cycles exhibiting a sawtooth pattern (Figure 2), the most characteristic climate feature of the last ~800 kyr. The 100-kyr cycle exhibits the strongest climate signal during the late Quaternary but corresponds to the eccentricity cycle, variations of which are known to produce the weakest

---

Plate 1. Correlations between (A) GISP2 δ^{18}O record [Grootes et al., 1993, Grootes and Stuiver, 1997; Stuiver et al., 1995; Stuiver and Grootes, 2000; chronology of Meese et al., 1994 and Sowers et al., 1993] and (B to G) planktonic and benthic foraminiferal δ^{18}O and δ^{13}C records from Ocean Drilling Program Site 893, Santa Barbara Basin, California, for past 60 kyr including (B) thermocline-dwelling *Neogloboquadrina pachyderma* δ^{18}O (blue), (C) surface-dwelling *Globigerina bulloides* δ^{18}O (red) [Hendy and Kennett, 1999; Hendy et al., 2002], (D) benthic δ^{18}O as a five-channel binomial average (green), (E) benthic δ^{13}C record (species shown near center of figure), and planktonic δ^{13}C records for (F) *G. bulloides* (red), and (G) *N. pachyderma* (blue) [Kennett et al., 2000a]. The warm Dansgaard-Oeschger (DO) interstadials (numbered 1 to 17) [Dansgaard et al., 1993] are clearly shown punctuating the otherwise cold conditions (stadials and glacials) of the last glacial episode. Shading represents intervals of laminated sediments [Behl and Kennett, 1996] associated with warming (interstadials and Holocene). Note benthic δ^{13}C oscillations associated with millennial-scale climate change and brief, episodic negative excursions in planktonic and benthic δ^{13}C records. ODP Site 893 chronology based on radiocarbon ages and stadial-interstadial transition tie points to GISP2 ice core [Hendy and Kennett, 2000].

orbital effects on incident solar radiation (maximum insolation change from eccentricity is only 0.2%). Therefore, the dominant 100-kyr climate cycle and associated sawtooth pattern of climate behavior require explanation.

It appears to take the entire build-up phase of each glacial cycle to produce ice sheets and associated sea-level lowstands of sufficient magnitude to somehow reset the sensitivity of the system to its interglacial mode [Denton et al., 1999]. Once reset, major warming events at the terminations could be triggered by the small effects of eccentricity on the amplitude of precession and half-precession insolation. In other words, the feedbacks that drove late Quaternary glacial terminations were established when ice sheets were near maximum extent. Denton et al. [1999] have suggested that the existence of large ice sheets was a necessary condition for the initiation of the last glacial termination.

Millennial-Scale Climate Variability

Pervasive millennial-scale climate oscillations are now well known to have persisted during the latest Quaternary (Figures 1 & 3; Plate 1) [Sarnthein et al., 2000], superimposed on Milankovitch-band insolation cycles (including glacial-interglacial episodes). These oscillations are recognized in ice cores, terrestrial lakes and bogs, speleothems, and marine sediments from the North Atlantic, Pacific and Indian Oceans [Dansgaard et al., 1993; Porter and An, 1995; O'Brien et al., 1995; Kotilainen and Shackleton, 1995; Kennett and Ingram, 1995a, b; Thunell and Mortyn, 1995; Lowell et al., 1995; Charles et al., 1996; Curry and Oppo, 1997; van Geen et al., 1996; Lund and Mix, 1998; Bond et al., 1997; Arz et al., 1998; Sachs and Lehman, 1999; Hendy and Kennett, 1999, 2000; Hendy et al., 2002; Peterson et al., 2000; Grigg et al., 2001; Kanfoush et al., 2000; Y. Wang et al., 2001; Altabet et al., 2002; Voelker et al., 2002; Sarnthein et al., in press; Kennett and Peterson, 2002]. Evidence exists for strong involvement of the polar, mid-latitude, and tropical regions in this change. First identified in Greenland ice cores, a series of warm interstadials (200 to 2500 yr durations) (Figure 1) punctuated the otherwise cold conditions of the last glacial episode [Dansgaard et al., 1993; Bond et al., 1993; Sarnthein et al., in press]. These climatic oscillations are often referred to as the Dansgaard-Oeschger (D/O) cycles [Dansgaard et al., 1984; Oeschger et al., 1984]. The air temperature shifts between these changes in climate state were large, likely ~6° to 10°C [Broecker, 2000] and remarkably abrupt, within decades to years. Thus these climate oscillations essentially reflect major, abrupt shifts in the ocean/atmosphere system between cool and warm states [Ditlevsen et al., 1996; Broecker, 1995; Paillard, 2001] associated with changes in greenhouse gas composition and albedo. Stadial-interstadial cycles are recorded by changes in the $\delta^{18}O$ in polar ice (Figure 1), $\delta^{18}O$ in foraminiferal tests in marine sediments (Plate 1; Figure 3), pollen assemblages in lake sediments [Allen et al., 1999; Grigg et al., 2001], and $\delta^{18}O$ in speleothems [Y. Wang et al., 2001]. These climate events are also asso-

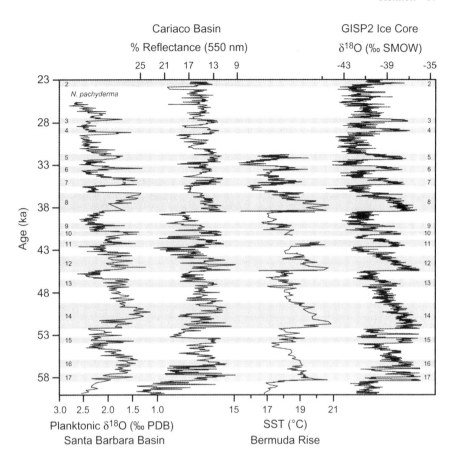

Figure 3. Comparison of high-resolution records of rapid, millennial-scale climate oscillations (Dansgaard-Oeschger (D/O) cycles) from distinctly different regions during the last glacial episode between 60 and 23 ka [Kennett and Peterson, 2002] including (from left to right) *N. pachyderma* $\delta^{18}O$ from ODP Site 893, Santa Barbara Basin, California [Hendy and Kennett, 1999], measured color reflectance (550 nm) of sediments from ODP Hole 1002C, Cariaco Basin, Venezuela, interpreted to reflect changing near-surface ocean productivity [Peterson et al., 2000], alkenone-derived sea-surface temperature estimates from core MD 95-2036, Bermuda Rise [Sachs and Lehman, 1999], and GISP2 $\delta^{18}O$, Greenland ice core [Grootes et al., 1993, Grootes and Stuiver, 1997; Stuiver et al., 1995; Stuiver and Grootes, 2000]. The records exhibit interstadials 2 through 17 (shaded) [Dansgaard et al., 1993].

ciated with significant changes in trace atmospheric gases (CH_4, CO_2, N_2O) (Figures 4 & 5) and in the marine and terrestrial biospheres [e.g., Cannariato et al., 1999; Cannariato and Kennett, 1999].

The discovery of millennial climate variability was instrumental in forcing reevaluation of the processes driving the late Quaternary climate system, because

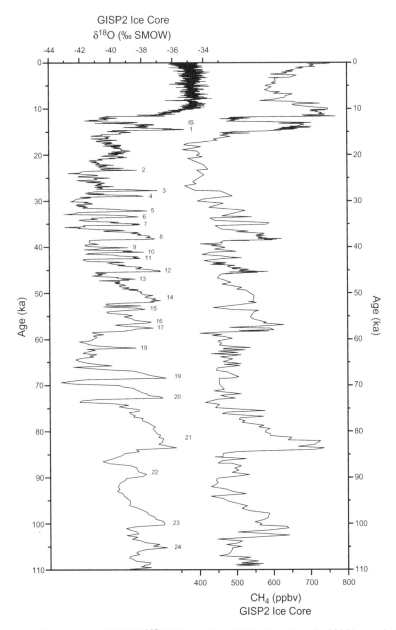

Figure 4. Comparison of GISP2 $\delta^{18}O$ (Figure 1) and CH_4 [Brook et al., 2000] records for the last 110 kyr from the Greenland ice core reveal a close correspondence of millennial-scale oscillations in air temperatures over Greenland and atmospheric CH_4 concentrations. Interstadials 1 to 24 are present [Dansgaard et al., 1993]. Chronology of $\delta^{18}O$ record same as Figure 1. The GISP2 gas-age time scale from Brook et al. [1996].

CH$_4$ (ppbv)

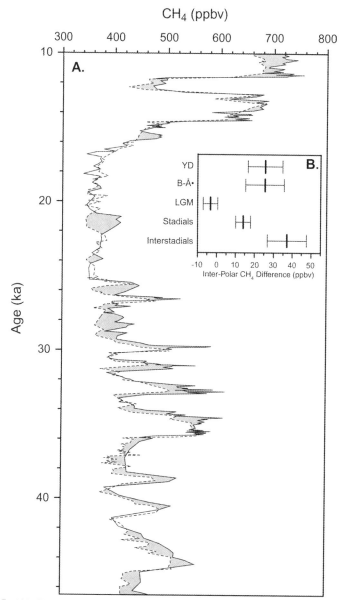

Figure 5. (A) Comparison of mean CH$_4$ concentrations from Greenland (GRIP) and Antarctica (Byrd and Vostok) ice core records between 46 and 10 ka. Antarctica: dashed line. Greenland: solid line. Intervals where Greenland CH$_4$ concentrations are greater than Antarctica are shaded. (B) Inter-polar CH$_4$ difference and 1σ error bars for five intervals: warm (interstadial) and cold (stadial) episodes during the last glacial episode, the Last Glacial Maximum (LGM), the Bølling-Ållerød (B-Å) and the Younger Dryas (YD) episodes [after Dällenbach et al., 2000, and references therein].

no obvious external forcing exists at these frequencies. Although there is currently no consensus for the origin of the abrupt climate changes, their existence suggests that internal feedback mechanisms, including greenhouse gas variations, played a crucial role both in setting the sensitivity of the climate system and creating or amplifying major change at orbital and sub-orbital time scales. Understanding what the amplifiers are and how the feedback mechanisms operate is central to understanding the character of late Quaternary climate change. The two most frequently considered hypotheses are that rapid millennial-scale cycles were triggered by (1) changes in thermohaline circulation associated with North Atlantic Deep Water (NADW) production or (2) changes in tropical heat distribution. Although these processes appear to have played important roles in the climate oscillations, computer models used to test the leading hypotheses suggest that it is unlikely that they could have, by themselves, caused such large and abrupt climate changes as recorded or maintained the warmth of the interstadials. Models have so far failed to fully explain the climate record including the magnitude, speed and extent of the climate changes during the shifts [National Research Council, 2002].

The thermohaline circulation hypothesis links changes in surface-water salinity at high northern latitudes of the Atlantic with abrupt switches in the strength of North Atlantic thermohaline circulation ("The Great Ocean Conveyer" [Broecker, 1999]). This in turn severely affects climate of the North Atlantic [Broecker et al., 1985; Manabe and Stouffer, 1997; Keigwin and Boyle, 1999; van Kreveld et al., 2000; Sarnthein et al., in press; Clark et al., 2002]. During glaciations, the circum-North Atlantic ice sheets provided a ready supply of freshwater to the North Atlantic, and perhaps played a key role in causing large, abrupt climate change through iceberg release, rerouting of continental drainage pathways, and episodic drainage of glacier-dammed lakes [Clark and Mix, 2000; Clark et al., 2001]. Thus, Termination 1A is explained by the sudden development of strong conveyor circulation followed by weak circulation because of low salinity surface waters associated with major glacial melting. This inhibited downwelling that diminished thermohaline circulation (NADW formation) at the onset of the Younger Dryas [Broecker 1990; Broecker et al., 1990] and decreased heat transport from the equator to the high latitudes. In this hypothesis, ice sheet dynamics are central in controlling climate behavior [Berger and Jansen, 1994; Bond et al., 1992, 1993; Paillard and Labeyrie, 1994]. Furthermore, Heinrich Events (represented by Heinrich Layers [Heinrich, 1988]) were associated with meltwater episodes at high North Atlantic latitudes, including massive iceberg discharge related to breakup of the adjacent ice sheet edges immediately prior to the rapid warming shifts [Seidov and Maslin, 1999; van Kreveld et al., 2000]. Heinrich Events (see Broecker and Hemming [2001]) occurred every 7 to 13 kyr during the last glacial episode, were brief (100 to 500 yr) [Dowdeswell et al., 1995], and are represented by sediment layers that are especially thick (several meters) in the Northwest Atlantic and become thinner to the east [Grousset et al., 1993]. Outside of the North Atlantic, Heinrich Events are often represented only as a second-order modulation of the timing and amplitude of the D/O cycles [Bond et al., 1993]. Nevertheless, their association with collapse of the deep-water conveyor system reflects the importance of these events in global climate change [Seidov and Maslin, 1999; Broecker and Hemming, 2001]. Following Heinrich Events, a decrease in melt-water discharge reactivated NADW production, increasing advection from the tropics and contributing to climate warming [Bond et al., 1992, 1993; Paillard and Labeyrie, 1994]. Although strong evidence supports such a sequence of events [van Kreveld et al., 2000], the hypothesis appears to be insufficient to explain the abruptness, magnitude, and global extent of the warmings. The largest problem with this hypothesis is its limitations in affecting rapid climate change beyond the North Atlantic region [National Research Council, 2002]. As documented later, evidence now exists that sources of upper intermediate waters also oscillated in close concert with millennial climate changes (Plate 1) [Hendy and Kennett, submitted; Kennett et al., 2000a].

The second hypothesis for the driver of millennial-scale climate change relates to changes in tropical heat distribution and consequent modulation of atmospheric water vapor. At present, atmospheric water vapor, the most important greenhouse gas, is dominantly produced in the tropics; its atmospheric abundance depending exponentially on temperature [Oltmans and Hofmann, 1995]. Thus, it can exert strong positive feedback on climate change and has rapid response times [Pierrehumbert, 1999]. Tropical surface ocean temperature changes in phase with higher latitude air and ocean temperature records suggest that the tropics have not passively responded to climate change triggered by other mechanisms such as changes in thermohaline circulation [Bard et al., 1997]. Yet several modeling studies suggested the opposite—that tropical climate and associated heat distribution responded to changes in North Atlantic thermohaline circulation [Fawcett et al., 1997]. Although it is clear that the tropics were strongly involved in global climate change [Cane and Evans, 2000; Peterson et al., 2000; Kienast et al., 2001; Y. Wang et al., 2001], the nature of the forcing agent for tropical change remains undetermined.

Any proposed mechanism for late Quaternary climate change must be consistent in explaining patterns of change in the earth system on orbital, millennial, and decadal time scales, as well as their global or regional distribution. For example, it is necessary to explain why, during deglaciation, a major reduction in ice volume followed each maximum in ice sheet expansion [Denton et al., 1999; Alley and Clark, 1999]. Stadial-interstadial (D/O) cycles are associated with rapid changes in precipitation and continental runoff from northern South America [Arz et al., 1998; Peterson et al., 2000] associated with oscillations in the position of the Intertropical Convergence Zone. High fluxes of continental sediment correspond to episodes of increased rainfall and are in phase with interstadials. Stadial-interstadial cycles are also recorded in southeast Asia (South China and Japan Seas), caused by increased summer monsoon strength during interstadials [Wang and Oba, 1998; Wang et al., 1999a] and monsoon-related wet and dry cycles in central and east Asia [Xiao et al., 1999; Y. Wang et al., 2001] that also affected changes in Japan Sea salinity [Wang and Oba, 1998; Tada et al., 1999]. The climate cycles are also recorded as denitrification and productivity signals in the Arabian Sea in response to modulation of the intensity of summer monsoonal upwelling [Altabet et al., 2002]. Understanding the causes of these rapid climate cycles and their interaction with orbital forcing is key to understanding late Quaternary climate variability. The rapidity of the climate changes indicates that the amplifying feedbacks were able to respond within years to decades. Curiously, this climate instability was most pronounced during the last glacial episode, at a time when large ice sheets led to the exposure of continental shelves.

Rapid temperature increases at stadial terminations are followed by general cooling during interstadials, and sudden cooling at interstadial terminations (Figures 1 & 3; Plate 1). This behavior imparts a sawtooth pattern to each stadial-interstadial cycle. Many stadial terminations are associated with brief temperature overshoots as recorded in the high-resolution climate records from Greenland [Dansgaard et al., 1993], Santa Barbara Basin [Hendy and Kennett, 1999] and the Sargasso Sea [Sachs and Lehman, 1999] (Figures 1 & 3; Plate 1). The especially high-resolution alkenone record from the Sargasso Sea [Sachs and Lehman, 1999] also exhibits major temperature instability (3° to 5°C) during stadial terminations. This high-resolution, short-term variability has not been documented in Greenland ice cores but may have occurred within the stadial terminations, intervals of less than 250 yr [Sachs and Lehman, 1999]. Similar SST variability is also exhibited during the longer interstadials recorded in Santa Barbara Basin [Hendy and Kennett, 1999] (Plate 1). These records indicate that stadial terminations were not uniformly smooth, but marked by large-scale variability.

Late Quaternary climate cycles form a pervasive 1 to 2 kyr climate rhythm with approximate spacing of ~1400 to 1500 yr that continues through the Holocene [Bond et al., 1997, 1999; Grootes and Stuvier, 1997; Mayewski et al., 1997; Sarnthein et al., in press]. During the last glacial episode, this 1500-yr cycle also occurred at high latitudes (42°S) in the South Atlantic as evident in records of ice-rafted sediment derived from the Antarctic Peninsula [Bond, 2000; Kanfoush et al., 2000] suggesting bi-hemispheric climate behavior. Also apparent is a distinct ~7.4-kyr oscillation with each episode consisting of several stadial-interstadial cycles of decreasing duration and amplitude (Figure 1) [Bond et al., 1993], thus, exhibiting a sawtooth pattern. These are the so-called Bond Cycles, each reaching a climax with a severely cold Heinrich Event marked by an interval of North Atlantic ice-rafting [Heinrich, 1988; Bond et al., 1993].

Late Quaternary climate behavior was therefore marked by a pervasive sawtooth pattern evident in individual stadial-interstadial cycles, Bond Cycles, and orbital cycles, including the dominant 100-kyr cycle (Figures 1-3; Plate 1). It is still unknown if this climate variability represents true periodicity [Bond et al., 1997], quasi-periodicity, or if it is fractal resulting from self-organized criticality [Bak, 1996]. Wunsch [2000] has even suggested that it resulted from aliasing.

Nevertheless, the millennial-scale rhythm appears to set the timing of the abrupt warming steps, with their magnitude in part modulated by their position within the 100-kyr cycle and Bond Cycles. McManus et al. [1999] suggested that the amplitude of millennial-scale climate response is controlled by a baseline climate state (ice volume threshold). They suggested that sensitivity increased during the last glacial episode when sea level was 30 m lower than present, but decreased during the LGM when sea level was more than 90 m lower. As a result, temperature variability, reflected as stadial-interstadial oscillations, was largest at intermediate glacial extent, and lowest during the Holocene and LGM (Figure 1; Plate 1). The enhanced climate variability within this band was not directly forced by ice sheet instabilities, but instead arose through modulation of the 1500-yr cycle [Bond et al., 1999]. Alley et al. [2001] and Rahmstorf and Alley [2002] have suggested in their stochastic resonance hypothesis that the mode switches resulted from weak periodic forcing (~1500 yr) combined with "noise" from ice sheet-related events. In this hypothesis, the climate switches causing the millennial-scale climate variability during the last glacial episode were triggered by more or less random events when the phase of the periodic forcing was favorable. The nature of the initial forcing mechanism remains unknown but suggested mechanisms summarized by Bond et al. [1999] include climate responses at the harmonic frequencies of orbital cycles, internally forced oscillatory behavior of the ocean/atmosphere system, and solar forcing. Additionally, Perry and Hsu [2000], Neff et al. [2001], and Bond et al. [2001] have suggested that a millennial-scale Holocene climate pulse resulted from solar variability. Also, Keeling and Whorf [1997, 2000] have suggested global oceanic tides as an underlying mechanism, whereby the surface ocean in key regions is periodically cooled by increased vertical water mass mixing affecting thermohaline circulation.

Although the changes in the earth system outlined above were not entirely synchronous, the near-synchrony (within the limits of dating) over broad areas of the Earth is striking [White and Steig, 1998; Steig et al., 1998] (Figure 3). Climate changes seem to be in phase throughout the Northern Hemisphere and appear to extend into southern temperate latitudes [Lowell et al., 1995; Denton and Hendy, 1994; Moreno et al., 2001; Kanfoush et al., 2000]. Paleoclimatic changes since the LGM are considered synchronous between North and South America [Whitlock et al., 2001; Behling and Hooghiemstra, 2001; Bradbury et al., 2001; Ledru et al., 2002], although the extent of ice cover was vastly different and changes in precipitation were not necessarily in the same direction [Bradbury et al., 2001]. This apparent synchroneity occurred in spite of contrasting large scale interhemispheric controls including major differences in extent of ice cover and opposing seasonal variations in insolation in the two hemispheres [Whitlock et al., 2001]. This suggests a primary role of the atmosphere (greenhouse gas content and albedo) rather than ocean circulation [Denton et al., 1999]. The question of synchronous climate change between the hemispheres, however, has been one of the most difficult to resolve. Records of atmospheric change and its direct effects, such as glacial extent and SST, strongly suggest synchroneity of the hemispheres. For example, the chronology and pattern of climate change from the Cariaco Basin, Santa Barbara Basin, Sargasso Sea (Figure 3), Iberian margin, Japan and South China Seas, Mainland China, and the equatorial Indian Ocean suggests in-phase response between Greenland, northern mid-latitudes, and equatorial regions of the oceans [Bard et al., 1997; Sachs and Lehman, 1999; Hughen et al., 1998, 2000; Peterson et al., 2000; Kennett and Ingram, 1995a, b; Ingram and Kennett, 1995; Hendy and Kennett, 2000; Hendy et al., 2002; Bender et al., 1999; Shackleton and Hall, in press; deMenocal et al., 2000; Y. Wang et al., 2001; Altabet et al., 2002]. Evidence also exists for tropical inter-hemispheric synchroneity during the last deglaciation from a southern Indian Ocean record (20°S) [Bard et al., 1997]. Furthermore, it has been suggested that interhemispheric synchroneity of millennial-scale climatic variability may also have extended to Southern Hemisphere latitudes as high as northern Antarctic waters [Kanfoush et al., 2000].

Compelling evidence also suggests synchronism of glacial discharge during the last glacial episode from the Northern Hemisphere ice sheets clustered around the North Atlantic (40° to 60°N) implying a common mechanism involved in Northern Hemisphere glacial activity [Elliot et al., 1998; Piper and Skene, 1998; Scourse et al., 1999]. Synchronous responses from several Northern Hemisphere ice sheets during Heinrich Events indicate a strong linkage between the various glacial systems, from small ice streams to the Laurentide ice sheet. This glacial behavior supports the interpretation that external forcing such as sea-level rise [Piper and Skene, 1998; Elliot et al., 1998; Scourse et al., 1999] and/or instabilities such as those associated with global warming, rather than internal ice-sheet behavior, triggered the events without a significant response lag. Multiple sources of reworked sediments associated with Heinrich Events [Piper and Skene, 1998; Bout-Roumazeilles et al., 1999] support this interpretation.

Apparent climate synchronism is perhaps best illustrated by the decadal-to-millennial-scale oscillations in alkenone paleotemperature records from subtropical, North Atlantic central gyral (Bermuda Rise, Sargasso Sea, Bahama Outer Ridge; ~28° to 34°N) sequences that co-vary with the Greenland ice core records [Sachs and Lehman, 1999] (Figure 3). In this region, temperature oscillations during the longest interstadials were large, comparable in magnitude to the LGM to Holocene change (3° to 5°C). Maximum interstadial SSTs (2° to 5°C warmer than stadials) are similar (within 0.5° to 1°C) between the three sites suggesting a broad regional signal. The magnitude of SST variations was 1/3 to 1/2 of Greenland air temperature variations; much larger than expected by Sachs and Lehman [1999] for a central gyral oceanic region (Figure 3). They interpreted their records to be instantaneously (within days to years) linked with Greenland via the atmosphere because of no known mechanism that might delay the climate signal between middle and high latitudes, and yet preserve the linear relation of amplitude for both brief and long lasting interstadials.

Evidence also suggests that snowlines changed symmetrically about the globe, at least between areas as distant as the Chilean Andes, New Zealand, European Alps, and North America [Denton et al., 1999]. A similar deglacial history is also suggested for South Africa [Abell and Plug, 2000]. This indicates not only that late Quaternary climate cycles at temperate southern latitudes were synchronous with those in the temperate northern latitudes, but also that the magnitudes of glacials and stadials were comparable between the two regions. Latest Quaternary ice advances and snowline depression at 40°S closely match those at 40°N [Denton et al., 1999] (although for an alternative view for the southern Chilean area see Bennett et al. [2000]). It appears likely that in high latitude late Quaternary sequences in the South Atlantic (including northern Antarctic waters), intervals marked by abundant ice-rafted sediments and associated maxima in benthic $\delta^{13}C$ (reflecting major increase in North Atlantic Deep Water production) correlate with the warm phases of the D/O cycles between 74 and 20 ka in the Greenland ice cores [Kanfoush et al., 2000]. This suggests an interhemispheric linkage that influenced the stability of the Antarctic ice sheet [Kanfoush et al., 2000] during the warmth of the D/O interstadials. Additional support for such inter-hemispheric synchronism of warm episodes within the Southern Ocean is suggested by inferred moisture source changes in the Dome C ice core (Antarctica) based on deuterium excess [Stenni et al., 2001].

Thus, there is much evidence indicating that climate change was synchronous about the equator, and the large size of the Northern Hemisphere ice sheets during the glacial episodes reflects more land surface available in the Northern Hemisphere than the Southern Hemisphere rather than greater cooling. Furthermore, $\delta^{18}O$ changes in tropical ice cores (5 to 6‰) between the LGM and Holocene were similar to those in Greenland, suggesting hemispherically similar temperature changes [Thompson et al., 1995, 1998]. These relations suggest greenhouse gas feedbacks were dominant during the late Quaternary. The pattern of climate change over the last glacial termination based on New Zealand pollen records provides strong evidence for synchronous inter-hemispheric climate changes within the limits of dating [Newnham and Lowe, 2000]. In contrast, evidence for possible antiphasing of ocean climate change in some areas of both Northern and Southern Hemisphere low to mid-latitudes [Rühlemann et al., 1999; Vidal et al., 1999; Kiefer et al., 2001] suggest an overriding local oceanic influence, perhaps in response to changes in thermohaline circulation.

The history of late Quaternary climate change in the Antarctic region may be in strong contrast to the synchronous climate changes propagated through the atmosphere over much of the globe. Climate records from high South Atlantic latitudes [Charles et al., 1996] and ice cores from interior Antarctica (Byrd and Vostok) suggest glacial and stadial terminations were antiphased with those in the Northern Hemisphere [Sowers and Bender, 1995; Bender et al., 1994; Shackleton, 2001]. Yet these differ from the in-phase near-coastal Taylor Dome ice core [Steig et al., 1998; Grootes et al., 2000]. Interstadials (between 75 and

35 ka) and Termination 1 recorded in Antarctica (Byrd and Vostok ice cores) preceded similar events in Greenland by as much as 1000 yr [Sowers and Bender, 1995; Blunier and Brook, 2001]. This analysis is in possible conflict with that of Bender et al. [1999] (supported by Hinnov et al. [2002]), who concluded, using the $\delta^{18}O$ of O_2 signal for inter-polar correlation, that both long and short interstadials had counterparts in both Greenland and Antarctic interstadials (but not Termination 1). These also appear to have been in phase, although uncertainties allow for corresponding events to be out of phase by as much as 1.4 kyr. Peak warmings were in phase in both polar regions. The general pattern and magnitude of the millennial-scale oscillations are thus similar between Greenland and Antarctica (Figure 1 in Blunier and Brook [2001]), although Antarctic interstadials were marked by more gradual warmings and coolings compared with the abrupt transitions in Greenland.

Melt-water pulse–1A event, at the onset of the abrupt warming that marked the beginning of the Bølling (Termination 1A), produced an ~20 m rise in sea level. Using regional sea-level changes to determine melt-water sources, Clark et al. [2002] suggested that Antarctica was an important source for this melt-water pulse, in addition to the Northern Hemisphere ice sheets. Melt-water pulse–1A occurred in less than 1000 yr, with much of the resulting sea-level rise occurring in 300 yr between 14.5 and 14.2 ka [Clark et al., 2002]. If, as suggested by Clark et al. [2002], major rapid melting of both Northern Hemisphere and Antarctic ice sheets contributed to this rapid and short-lived sea-level rise, then synchroneity, rather than phasing, of rapid warmings in the polar regions of both hemispheres would be indicated.

The exact relative timing of these events recorded in opposite polar regions is crucial to our understanding of late Quaternary climate behavior. Blunier and Brook [2001] considered it unlikely that the Greenland/Antarctic temperature offset is an artifact resulting from chronological limitations. Instead, they suggested that antiphased changes during interstadial climate episodes of various magnitudes between Antarctic climate and that further north may be explained by alternating influence of deep water formation in the north Atlantic and the Southern Ocean—the so-called "bipolar see-saw" [Broecker, 1998]. This hypothesis proposes that SSTs are antiphased between the high latitudes of the North and South Atlantic, because heat is transported by surface currents northwards across the equator to balance water lost during NADW production [Manabe and Stouffer, 1997]. Blunier and Brook [2001] expressed surprise that such a large apparent phase shift was maintained between 90 and 10 ka in spite of large changes in background state of the climate system including ice volume, sea level, and orbital geometry. Furthermore, it seems remarkable that the relative magnitudes and durations of individual and groups of interstadial events were maintained between both polar regions with a lag of 1000 yr or more.

The climate record from the Taylor Dome ice core is problematic when compared with records from other parts of the Antarctic. The pattern of magnitude and duration of stadial-interstadial cycles in the Taylor Dome ice core resembles that of Greenland, suggesting climate synchronism between the two areas. However, under full glacial climate (Marine Isotope Stage (MIS) 4, late MIS 3 and MIS 2) the Taylor Dome record is similar to that of Vostok [Grootes et al., 2000]. For the deglacial interval (last 28 kyr), distinct phasing is exhibited between several ice cores from East Antarctica [Jouzel et al., 2001]. Jouzel et al. [2001] have questioned the synchronicity of the deglacial warm events recorded in the Taylor Dome and Greenland ice cores, especially the interval before 14 ka. The primary differences between Greenland and Antarctica relate to the timing of the climate reversal during the last deglaciation. In Antarctica, the Antarctic Cold Reversal is considered to correlate with the Bølling-Ållerød in Greenland, an interval of extreme warmth. Furthermore, the Younger Dryas cooling has not been identified in Antarctica (except in Taylor Dome), and is considered to have occurred during the warming trend following the Antarctic Cold Reversal [Monnin et al., 2001]. In general, the records are considered to be in antiphase [Bender et al., 1994], although this clearly is more complex when the Taylor Dome record is considered. Interestingly, the most recent climate trends of Antarctica appear to be antiphased with global trends. Since 1966, the Antarctic continent has experienced net cooling (e.g., 0.7°C per decade in the McMurdo Dry Valleys) at the very time when average air temperatures at the Earth's surface have been increasing by ~0.06°C per decade during the 20th century and 0.19°C per decade from 1979 to 1998 [Doran et al., 2002]. This suggests that antiphasing of Antarctic climate can occur during times of global warming.

The climate records of Subantarctic Ocean sediment cores in the South Atlantic appear to support the see-saw model on orbital time scales, but not on millennial time scales [Ninnemann et al., 1999], which may be synchronized with the Northern Hemisphere. Broecker [2000] suggested that the boundary between the dominant northern pattern of climate behavior and the Antarctic pattern lies in the Southern Ocean. However, Jouzel et al. [2001] considered the see-saw model of climate control by oscillating strength of North Atlantic versus Antarctic deep water formation too simplistic. Atmospheric changes over all or part of Antarctica may be more strongly affected by Southern Ocean influences than by atmospheric changes to the north. Strong inertia of the Antarctic climate system should result from the strength of the circumpolar ocean/atmosphere system and have led to great uniformity of Antarctic climate during the Quaternary. Major spatial heterogeneity of Antarctic climate, if true, would be surprising.

Caution must be taken, however, when synchroneity or antiphase behavior is identified between high-resolution records from distant regions. At present, dating techniques of paleoclimatic sequences remain insufficient to provide conclusive proof of synchroneity or antiphase behavior. Layer counted records such as varved lake and marine sequences and ice cores contain errors produced by indistinguishable or doubled intervals, such that the GISP2 ice core record has dating errors of ~200 yr during the last termination [Grootes and Stuvier, 1997; Sarnthein et al., 2000]. Radiocarbon datums include inaccuracies related to instrumental error (50 to 200 yr depending on the amount of carbonate available for analysis), changes in past ^{14}C production, and poorly constrained surface-water reservoir ages for marine sequences. Calibration between calendar and ^{14}C years produces errors of up to 1000 yr during deglaciation and possibly higher during MIS 3 [Beck et al., 2001]. Paleomagnetic intensity records may eventually improve dating abilities during MIS 3 [Stoner et al., 2000; Channell et al., 2000], but will not distinguish synchroneity on the order of years. Consequently, synchroneity can only be demonstrated or refuted within the limits of our dating ability.

Although inconclusive and hindered by dating issues, an almost global pattern (other than Antarctica) of climate change appears to have occurred on orbital and sub-orbital time scales during the late Quaternary. If so, a globalizing mechanism has been instrumental in affecting this pattern of climate change. It seems clear that two major processes operated to create rapid changes on millennial and shorter time scales: (1) changes in greenhouse gas concentrations [Petit et al., 1999; Cuffey and Vimeux, 2001] and (2) rapid reorganizations of ocean circulation [Broecker et al., 1985]. In certain regions, ocean circulation changes exert a dominant control on local climate change compared with signals transmitted through the atmosphere. This appears to be particularly strong in the Atlantic Ocean where rapid changes in thermohaline circulation have major influence [Ninnemann et al., 1999]. In contrast, SST oscillations off southern California [Hendy and Kennett, 1999; Hendy et al., 2002] appear synchronous with those in the North Atlantic and consistent with essentially synchronous temperature change that would result from atmospheric forcing. It appears likely that the hemispheres were synchronized by changes in greenhouse gas composition as suggested by Petit et al. [1999]. Simple heat redistribution cannot alone account for the synchronous global cooling during the LGM or the apparently synchronous inter-hemispheric climate oscillations on various time scales north of the Antarctic region [Denton et al., 1999].

2

Atmospheric Methane Behavior: Ice Core Records

Trapped air in ice cores provides the most direct evidence for past changes in the atmospheric trace gases CH_4, CO_2, and N_2O [Flückiger et al., 1999] (see Raynaud et al. [1993, 2000] for reviews). In particular, these records have revealed remarkable changes in atmospheric CH_4 concentrations during the late Quaternary on orbital, millennial, and decadal time scales (Figures 2 & 4-6). Large, rapid variation in atmospheric CH_4 contrasts with much slower changes in atmospheric CO_2 (Figure 2). These records are of great significance for understanding late Quaternary climate dynamics because of the intimate relationship that exists between CH_4 and temperature records in ice cores. Determining the relationship between CH_4, CO_2, and climate at decadal resolution is of fundamental importance [Severinghaus and Brook, 1999]. We now summarize the major late Quaternary atmospheric CH_4 trends exhibited in ice cores and their association with climate events. These records pose two key questions: (1) What produced the rapid, large atmospheric CH_4 increases at glacial and stadial terminations? and (2) Were the large atmospheric CH_4 increases instrumental in forcing the rapid climate warming episodes that marked the late Quaternary?

Methane records preserved in ice cores are generally considered to represent a reliable record of past changes in atmospheric composition. Although atmospheric concentrations of CH_4 may have been higher than that indicated by concentrations in ice [Brook et al., 1996], rates of change appear to approximately reflect the rate of change in the atmosphere [Brook et al., 1996] (see "Future Tests of the Hypothesis" chapter). Interpretations of the source of atmospheric CH_4 variability during the late Quaternary must take into consideration: (1) the behavior, rates of change, and timing exhibited in ice core records; and (2) near-synchroneity between temperature and CH_4 variability, interpretations of which are complicated by age differences between the gas and the ice in which it is trapped because of their different accumulation mechanisms.

Methane Hydrates in Quaternary Climate Change
© 2003 by the American Geophysical Union
10.1029/054SP03

The Antarctic ice core records show that CH_4 and CO_2 exhibit remarkable similarities with Antarctic air temperatures throughout the late Quaternary [Petit et al., 1999] (Figure 2). The regularity of the CH_4 and CO_2 variations through several 100-kyr climate cycles (Figure 2) suggests a well-ordered set of dominant mechanisms [Sigman and Boyle, 2000] that are highly sensitized to change during the late Quaternary. One of the most significant observations made in ice core stratigraphy is of the intimate relationship between CH_4 and temperature oscillations in the pre-Holocene record of the late Quaternary [Jouzel et al., 1993; Brook et al., 1996; Petit et al., 1999]. Changes between the two climate parameters appear almost in lockstep (Figures 2 & 4), suggesting a common origin [Brook et al., 2000] (Figure 2) and a fundamental relation between CH_4 and temperature change. Petit et al. [1999] and Raynaud et al. [2000] have therefore suggested that greenhouse gases (CH_4 and CO_2) are important amplifiers of the initial orbital forcing of late Quaternary climate change and have significantly contributed to the glacial-interglacial oscillations. On the other hand, Severinghaus et al. [1998] and Brook et al. [1999] suggest little or no causal role for CH_4 in warming based on their calculation of phasing between CH_4 and temperature in Greenland ice cores. Clearly, if climate change was forced in part by CH_4 variations, the changes should be recorded in ice cores as synchronous or with a minor lag. However, as discussed below, determining this relationship in ice cores has been difficult because of the differences in age between the trapped gas and ice, variation in this age difference through time, and because the timing of CH_4 and temperature changes were extremely close. If CH_4 from methane hydrates played a role in the rapid warmings that marked the late Quaternary, increases in atmospheric CH_4 should have been synchronous or led the rapid warming episodes.

Oscillations on Orbital Time Scales

The Vostok ice core record, comprising the last 400 kyr [Petit et al., 1999], exhibits similar oscillations in temperature, CH_4, and CO_2 on orbital time scales (Figure 2), producing the late Quaternary sawtooth pattern of climate behavior. Each cycle commences with a dramatic warming marking the interglacial, followed by increasingly cooler episodes terminating with a glacial maximum. The next cycle commences with a rapid jump to the next interglacial. In most cycles, the coldest interval immediately precedes the rapid warming marking the glacial termination. The CH_4 and temperature changes exhibit remarkably close phase relations, each marked by rapid increases followed by slower decreases [Petit et al., 1999]. Furthermore, the magnitudes of CH_4 and temperature variations are very similar (Figure 2). In contrast, the CO_2 decreases clearly lag temperature decreases by several thousand years, although a more recent modeling study suggests a significantly decreased lag of only ~1000 yr, the approximate mixing time of the ocean [Cuffey and Vimeux, 2001]. Nevertheless, a general similarity exists between changes in temperature and CO_2, although this relationship does not hold up in detail (Figure 2) except during glacial terminations. Thus, temperature decreases during a glacial cycle are closely matched by CH_4 decreases, less so by CO_2 decreases.

Glacial Termination Jumps

The last four glacial terminations were accompanied by increases in atmospheric CH_4 from 320-350 ppbv to 650-770 ppbv and in CO_2 from ~180 to 280-300 ppmv (Figure 2). The increases in temperature, CH_4, and CO_2 were in phase at each of the glacial terminations [Petit et al., 1999]. Atmospheric CH_4 initially rose slowly, then jumped rapidly during the last termination [Monnin et al., 2001] (Figures 4 & 5). A small, initial CH_4 rise (~40 ppbv) began at ~17.5 ka, apparently synchronously with the warming episode (the Meiendorf Episode) that predates the B-Å [Chappellaz et al., 1993; Sowers and Bender, 1995; Brook et al., 1996; Bender et al., 1999]. At the Bølling (Termination 1A), the rapid CH_4 increase corresponds to the rapid warming in Greenland [Sowers and Bender, 1995; Bender et al., 1999; Petit et al., 1999; Monnin et al., 2001] marking the onset of major deglaciation. The earlier (~17.5 to 14.7 ka) CH_4 (and CO_2) increase may have contributed to warming and initial deglaciation in the Northern Hemisphere [Sowers and Bender, 1995]. However, the feedbacks that contributed to major warming and deglaciation at ~14.7 ka were significantly delayed, as recognized by Petit et al. [1999].

A high-resolution record of climate and greenhouse gas changes during the last 22 kyr from Dome Concordia, Antarctica [Monnin et al., 2001] found strong similarity between changes in CH_4 and temperature in the Greenland ice cores and Antarctic climate. General agreement also is evident between CH_4 and CO_2 changes, suggesting a linkage. However, similarities in the records disappear during the Younger Dryas when CH_4 values plummeted at a time of steady CO_2 increase. A general correspondence between CO_2 concentration and Antarctic temperature change suggests that the Southern Ocean played an important role in causing the CO_2 rise. Regardless of this observation, Monnin et al. [2001] believe similarities between the behavior of CH_4 and CO_2 indicate that CO_2 changes also resulted from processes related to the pre-Bølling warming that occurred outside of the Antarctic region.

High-resolution geochemical analyses [Severinghaus and Brook, 1999] show that Termination 1A was extraordinarily rapid with warming of $9°\pm3°C$ over several decades beginning at 14.672 ka. Severinghaus and Brook [1999] found that atmospheric CH_4 concentrations also rose abruptly (within ~50 yr) and suggest that the CH_4 increase began 20 to 30 yr after the onset of the abrupt warming. As discussed later, given the near-synchronism suggested by these experiments and within the context of much data summarized in this contribution, we have adopted the hypothesis that these changes were essentially synchronous (see "Future Tests of the Hypothesis" chapter).

The clear parallelism between changes in temperature and greenhouse gas concentrations have suggested to some a critical role for greenhouse forcing mechanisms in late Quaternary climate change [Raynaud et al., 1993; Petit et al., 1999; Monnin et al., 2001]. Petit et al. [1999] suggested that greenhouse gases caused 50% of the glacial-interglacial temperature increase. Calculations of the direct radiative forcing corresponding to the CO_2, CH_4, and N_2O changes during Termination IV (between MIS 10 and 9) suggest a temperature increase of ~1°C [Petit et al., 1999]. This initial forcing would have been reinforced by associated increases in relative humidity (water vapor), surface albedo (ice and vegetative cover), and planetary albedo (cloud cover). These workers concluded that greenhouse gas increase contributed significantly (2° to 3°C) to globally averaged glacial-interglacial temperature change. Each termination was marked by the same sequence of climate forcings: orbital forcing followed by two strong positive amplifiers, with greenhouse gases acting first, and then ice-albedo feedback associated with deglaciation.

Millennial-Scale Oscillations

An intimate relationship is also evident between millennial-scale CH_4 and temperature oscillations during MIS 3. High-resolution stratigraphy shows that all the interstadials, except possibly interstadial 20 [Brook et al., 1999] and 19 [Brook et al., 1996], are associated with CH_4 increases. Abrupt changes in CH_4 concentration of ~50 to 300 ppbv are coeval with interstadials [Brook et al., 1996]. In a well-studied example, Brook et al. [1999] found that the onset of interstadial 8 exhibits a 150 ppbv increase within only ~100 yr, thus rivaling the abruptness of the increase at the beginning of the Holocene (Termination 1B). The amplitude of the stadial-interstadial CH_4 variations is of the same magnitude in both the GRIP and Byrd ice cores [Stauffer et al., 1998] (Figure 5). The amplitude response between CH_4 and temperature is very similar, suggesting modulation by a related process.

The stadial episodes during MIS 3 were each marked by relatively low and stable atmospheric CH_4 concentrations (~450 ppbv) (Figures 4 & 5) [Brook et al., 1999]. Even during the extreme cold of the LGM, inferred atmospheric CH_4 values exhibited much the same low values (~350 ppbv) in spite of even more continental ice sheet growth and a maximum lowering of sea level. This suggests that the processes that produced CH_4 at times when concentrations were higher were already essentially turned off and that minimal variation in the balance of CH_4 sources and sinks occurred within the coldest intervals of MIS 4 through MIS 2. Thus CH_4-producing systems were relatively insensitive to even greater ice sheet expansion and lowering of sea level than had occurred during MIS 3 stadials. The abrupt warmings were associated with large changes in the CH_4 budget and thus reflect fundamental changes in the systems that produce and consume CH_4. Brook et al. [1999] considered that these low values during glacials and stadials reflect background levels of production in the absence of important wetlands. However, as discussed later, they may also reflect stability of the methane hydrate reservoir.

Compared to CH_4, CO_2 seems to have varied little from stadial to interstadial, but did change in association with Heinrich Events, especially those occurring before the longer interstadials. These relative variabilities also suggest a special role for CH_4 in the modulation of late Quaternary climate behavior. The in-phase relationship between temperature and CH_4 changes of similar magnitudes at both millennial and orbital time scales, suggests an active role for CH_4 in greenhouse forcing of late Quaternary climate.

Methane Overshoots

Dynamic relations between climate and CH_4 are also suggested by brief CH_4 "overshoots" associated with the rapid increases at glacial terminations. This behavior is especially conspicuous at Termination IV (330 ka) (Figure 2). These short-lived overshoots appear to have been associated with similarly brief increases in temperature (Figure 3; Plate 1). A conspicuous, brief CH_4 overshoot is also associated with Termination 1B at ~11.5 ka (Figure 6) [Blunier et al., 1998], which Blunier [2000] has suggested may have resulted from catastrophic CH_4 release from methane hydrates. This event also appears to be associated with a brief temperature overshoot (Plate 1). The onset of the Holocene thus exhibits the highest CH_4 values measured for the last 11.5 ka [Blunier et al., 1998] before anthropogenic contributions. The overshoots suggest a dynamic production of CH_4 rather than passive CH_4 increase resulting from ecological equilibration with climate warming.

Holocene Record

Atmospheric CH_4 remained relatively high (>550 to 775 ppbv) throughout the Holocene following the rapid rise at its beginning (Termination 1B) (Figures 4 & 6). However, a distinct pattern of change did occur through the Holocene [Blunier et al., 1995; Blunier, 2000; Brook et al., 1999; Raynaud et al., 2000] (Figure 6). The earliest 2 kyr of the Holocene was marked by the highest concentration (Greenland; ~725 to 750 ppbv) followed by a general decrease of ~150 ppbv to lowest concentrations during the mid-Holocene, with lowest values at ~5.5 ka. After the mid-Holocene, concentrations steadily rose to reach the relatively high concentrations (Greenland; ~725 ppbv) marking the last millen-

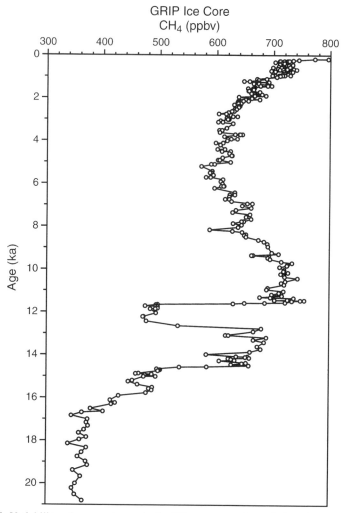

Figure 6. Variability of mean atmospheric CH_4 concentrations during the last 21 kyr recorded in the Greenland (GRIP) ice core [after Dällenbach et al., 2000, and references therein]. The plot ends at 800 ppbv, whereas the modern concentration is ~1,700 ppbv. Note distinct reduction of atmospheric CH_4 concentrations associated with the Younger Dryas centered at ~12 ka.

nium [Blunier et al., 1995] (Figure 6). A general decoupling of CH_4 and temperature is evident during the Holocene [Brook et al., 1996], although a sharp decrease of CH_4 and temperature occurred together at ~8.2 ka (Figure 6) [Blunier et al., 1995; Alley et al., 1997; Raynaud et al., 2000]. Prior to 7 ka, CH_4 concentrations exhibit greater variability (Figures 4 & 6). After 7 ka, changes were much smoother lacking the brief fluctuations that marked the early Holocene. Holocene CH_4 variations have previously been attributed to variations in the hydrologic cycle as it affected wetland development; mid-Holocene low concentrations linked to the desiccation of tropical lakes and late Holocene high concentrations with increasing contributions from northern high latitude wetland sources [Blunier et al., 1995].

Methane Oscillations: Sources or Sinks?

The changes in atmospheric CH_4 concentrations recorded in the polar ice cores theoretically could have resulted from changes in either the sources or sinks of CH_4. Most CH_4 emitted to the atmosphere is rapidly oxidized to CO_2 by OH radicals within the troposphere [Prinn et al., 1995]. Atmospheric chemical models developed to estimate tropospheric OH concentration changes for the late Quaternary indicate <17% change between glacial and interglacial episodes [Thompson et al., 1993]. Because glacial-interglacial atmospheric CH_4 concentrations doubled, the primary cause of the CH_4 variations must have resulted from changes in the sources rather than the sinks.

3

Methanogenesis and Methanotrophy

Methanogenesis

Biogenic methane is produced almost exclusively by a group of primitive anaerobic microbes, the methanogens (CH_4-producing Archaea), as an end product of their energy metabolism (see reviews in Rogers and Whitman [1991] and Ferry [1993]). Methane production, or methanogenesis, takes place only under anoxic reducing conditions. Methanogenic microbes produce huge quantities of CH_4 in a range of aquatic, anaerobic environments. Such environments include anoxic freshwater sediments from lakes, marshes, mires, floodplains, rice paddy fields and tundra peatlands, as well as marine sediments in a range of environmental settings where organic carbon is in sufficient concentrations. Methanogens play a major role in the anaerobic biodegradation of organic material.

Methanogenesis occurs exclusively in environments protected from the influence of O_2, which is toxic to methanogens [Boone, 1991]. In these settings, oxygen-deficient zones develop due to O_2 consumption by respiration and limitation of O_2 diffusion from the atmosphere or ocean [Conrad, 1989; Large, 1983]. Return of oxygen can rapidly shut off CH_4 production, such as occurs during the draining of wetlands. Methanogens also live in anoxic gastro-intestinal habitats of some mammals (ruminant artiodactyls) such as cattle and sheep and also in termites.

Methanogenesis occurs via several metabolic pathways, chiefly by CO_2 reduction or acetate fermentation.

$$CO_2 + 4H_2 \rightarrow CH_4 + 2H_2O \ (CO_2 \ \text{reduction})$$

$$CH_3COOH \rightarrow CH_4 + CO_2 \ (\text{acetate fermentation})$$

Methane Hydrates in Quaternary Climate Change
© 2003 by the American Geophysical Union
10.1029/054SP04

Most methanogens are thought to obtain their energy from the anaerobic oxidation of H_2 using CO_2 as the electron receptor [Large, 1983]. The H_2 is available as a byproduct of fermentation by other microbial species.

Methane is not directly produced from complex organic compounds, but is instead produced from a limited number of simple compounds themselves produced during the breakdown (fermentation) of complex organics. The structure of a methanogenic microbial community includes the bacteria that ferment complex organic molecules in addition to the methanogens that produce CH_4 directly from the simple fermentation products. In this process, a high diversity of microbial species contribute to the fermentation by degrading complex organic molecules to simpler compounds [Conrad, 1989], ultimately producing H_2, CO_2 and acetate which can be utilized by methanogens to form CH_4 [Conrad, 1989]. Most methanogens reduce CO_2 to CH_4 using the electrons derived by oxidizing H_2 or formate, or sometimes alcohols. Other methanogens utilize methyl groups such as acetate, methylated amines or methanol as the substrate for methanogenesis [Boone, 1991]. The removal of H_2 by methanogens is important (for thermodynamic reasons) for the continuation of fermentation by these other microbes, because increases in H_2 can cause fermentation to shut down. By living in close association with fermenters, methanogens maintain their substrate supply by utilizing excess H_2 in CH_4 production [Hoehler et al., 1998; Schimel, 2001]. Furthermore, the removal through methanogenesis of the major catabolic product of fermentation, acetic acid, not only benefits the energetics of its catabolic reactions, but also prevents accumulation of acidic products that could inhibit growth by lowering the pH [Conrad, 1989].

Environmental controls on CH_4 production can be complex but are associated with the temperature and pH of the environment, the presence of alternative electron acceptors and substrate quality. Substrate quality is ultimately controlled by the organic substrates entering the sediment and the environment of methanogenesis [Schimel, 2001].

The release of CH_4 from a system is dependent upon the balance between CH_4 production and consumption [Reeburgh et al., 1991]. Methane is consumed aerobically by microbes using it as their sole carbon and energy sources. These bacteria are the methanotrophs. They are crucial in their utilization of O_2 to recycle CH_4 into organic compounds thus making it available, as CO_2, to autotrophs [Large, 1983]. This completes the CH_4 cycle.

Not all anaerobic environments are suitable for methanogens. In the presence of high concentrations of sulfate, oxidized metals (Fe(III) or Mn(IV)) or oxidized nitrogen compounds, other anaerobic microbes dominate [Large, 1983]. Under these conditions, electron acceptors such as sulfate or ferric iron stimulate activity of other microbes, and inhibit methanogenesis. In marine environments where the concentrations of sulfate are high, sulfate-reducing bacteria will out compete methanogens for available substrate (H_2, acetate) and hydrogen sulfide production predominates over methanogenesis [Jones, 1991; Hoehler et al., 1998].

Organic carbon is oxidized by sulfate-reducing bacteria to CO_2 (depleting available acetate, formate and H_2) as sulfate is reduced to hydrogen sulfide. In marine sediments, this continues until the majority of seawater sulfate is reduced in pore waters by sulfate-reducing bacteria. Once sulfate is depleted the consumption of organic material becomes coupled to CH_4 production. High concentrations of sulfate in seawater severely limit CH_4 production in environments such as coastal mangroves, salt marshes, estuaries [e.g., Kennett and Hargraves, 1985] and marine surface sediments. In marine sediments, methanogens are largely restricted to depths below the sulfate reduction zone. Depending on bottom-water conditions and organic matter content of sediments, methanogenesis can occur from immediately below the sea floor to hundreds of meters into the sediment. Methanogenic bacteria can be active at great depths [Zengler et al., 1999; Parkes et al., 2000; Marchesi et al., 2001], producing large volumes of CH_4 that may be transformed into methane hydrate when pressure, temperature, and concentrations are suitable.

Methanotrophy

Methanotrophy is an important component of the global CH_4 system involving the interception and consumption of CH_4 by microbial communities [Hanson and Hanson, 1996]. This process is performed by anaerobic methanotrophs in anoxic sediments and waters and by aerobic methanotrophs in the water column. Methanotrophy is of major climatic importance because it results in a large reduction in CH_4 that otherwise would be transported to the ocean/atmosphere system. Although large volumes of CH_4 are stored in marine sediments, the ocean contributes only ~2% of the modern global flux of CH_4 to the atmosphere [Cicerone and Oremland, 1988].

Methane produced in marine sediments diffuses upwards often to be consumed under anoxic conditions by methanotrophs through a process in which sulfate acts as the terminal oxidant [Valentine and Reeburgh, 2000]. The anaerobic oxidation of CH_4 (for recent reviews see Valentine and Reeburgh [2000] and Valentine [in press]) requires both sulfate and CH_4 and takes place in anoxic environments including marine sediments, CH_4 seeps, anoxic waters and near methane hydrates [Alperin and Reeburgh, 1984; Valentine and Reeburgh, 2000; Zhang et al., 2002]. Methane oxidation provides energy needed for growth and metabolism of microbial communities. Estimates of the net global rate of anaerobic CH_4 oxidation vary widely from 70 Tg yr^{-1} (1 Tg = 10^{12} g) [Reeburgh, 1996; Reeburgh et al., 1993] to 300 Tg yr^{-1} [Hinrichs and Boetius, in press], equivalent to 12 to 55% of the net modern atmospheric CH_4 flux.

Anaerobic CH_4 oxidation [Barnes and Goldberg, 1976; Reeburgh, 1976; Martens and Berner, 1977] in most CH_4-containing marine sediments occurs at the base of the sulfate reducing zone, where upward fluxes of CH_4 encounter

SO_4^{2-} (the so-called CH_4 to sulfate transition). Oxidation of CH_4 is nearly complete in this zone and occurs through the following chemical reaction:

$$CH_4 + SO_4^{2-} \rightarrow HCO_3^- + HS^- + H_2O$$

Several factors can affect the depth at which anaerobic oxidation of CH_4 occurs in marine sediments including organic content, sedimentation rate, upward flux rate of CH_4 and downward flux rate of SO_4^{2-}, temperature and pressure. However in most marine sediments the most significant factors are organic content and CH_4 supply. The process occurs in a wide range of anoxic conditions even under extreme conditions of high sulfide levels, near-freezing temperatures and low energy conditions [Valentine, in press]. Anaerobic oxidation of CH_4 also occurs near methane hydrate deposits where CH_4 is dissolved in pore fluids, although hydrates themselves support few microbes [Lanoil et al., 2001]. The process also occurs in various sulfate-containing, anoxic waters such as the Black Sea and Cariaco Basin [Reeburgh, 1976; Reeburgh et al., 1991].

Anaerobic oxidation of CH_4 consumes sulfate and CH_4 to produce carbonate and hydrogen sulfide through the following reaction:

$$CH_4 + SO_4^{2-} \rightarrow CO_3^{2-} + H_2S + H_2O$$

This reaction has the effect of increasing the alkalinity of pore water which leads to precipitation of authigenic carbonate minerals [Roberts and Aharon, 1994].

The identities of the microbial communities catalyzing this reaction, as well as the underlying mechanisms are both areas of active investigation. The process of CH_4 oxidation appears to be performed in some environments by a consortium of CH_4-oxidizing Archaea and sulfate-reducing bacteria acting in syntrophic association [Hoehler et al., 1994; Valentine and Reeburgh, 2000]. The metabolism of these consortia appears to involve interspecies electron transfer, whereby the archael member of a consortium oxidizes CH_4 and transfers reduced compounds (possibly H_2 and acetic acid) to the sulfate reducing bacteria [Valentine and Reeburgh, 2000]. Continued CH_4 oxidation is presumably dependent on the constant consumption of intermediates, otherwise CH_4 oxidation is thermodynamically inhibited. This cooperative situation allows for CH_4 oxidation through interspecies electron transfer. There is also evidence that CH_4 oxidation may be performed by a single species of Archaea in some environments, though little is currently known about this organism.

Methanotrophy also occurs as a major and climatically important process in the ocean's water column. Most CH_4 entering the water column is dissolved within a few hundred meters above the ocean floor [Guinasso and Schink, 1973], except almost certainly in cases of dynamic (catastrophic) upward transport of CH_4. Dissolved CH_4 is further subject to oxidation by aerobic methanotrophs when the water column is oxic. The process represents an additional major barrier for transfer of CH_4 into the atmosphere and generally accounts for low CH_4 concentrations in the marine water column. Aerobic CH_4 oxidation produces CO_2 in the following reaction:

$$CH_4 + 2O_2 \rightarrow CO_2 + 2H_2O$$

The generation of CO_2 in water produces a weak acid that tends to dissolve carbonates, in contrast to anaerobic oxidation of CH_4 [Valentine, in press].

Aerobic methanotrophy is known to occur in marine waters, though little is known about the microbial communities responsible. Measurements by Valentine et al. [2001] suggest substantial methanotrophic activity in CH_4 released from the sea floor in Eel River Basin, northern California. Oxidation rates of CH_4 in plumes are relatively high suggesting active methanotrophy soon after release of CH_4 from the sediments. Turnover of CH_4 in deep waters with high concentrations of CH_4 were both constant and rapid (turnover time ~1.5 yr). However, in areas of low CH_4 concentrations, CH_4 consumption continues, although at much slower rates leading to turnover times of ~50 yr [Scranton and Brewer, 1978; Rehder et al., 1999]. These trends have suggested to Valentine [in press] that the active populations that mark areas of high CH_4 have not developed in areas of low CH_4 concentrations.

4

Source of Methane During Rapid Increases

The Prevailing Hypothesis: A Continental Wetland Source

The rapid atmospheric CH_4 increases during the late Quaternary are closely associated with intervals of rapid warming. Especially large, rapid CH_4 increases occurred at glacial terminations. During Termination 1A, CH_4 concentrations essentially doubled from LGM values (350 ppbv) approaching concentrations typical of the late Holocene (~700 ppbv) (Figures 2, 4 & 6). This large CH_4 increase occurred in less than a few decades [Severinghaus and Brook, 1999]. It is crucial to determine the source of such CH_4 increases in order to understand the processes forcing the rapid late Quaternary warmings.

Raynaud and Siegenthaler [1993] recognized that CH_4 began to increase before any significant melting of continental ice sheets at the rapid glacial terminations. Therefore any resulting radiative forcing that caused initial ice sheet melting did not result from sea-level change [Raynaud and Siegenthaler, 1993]. Furthermore, as CH_4 reached interglacial levels within a few decades during Termination 1A, ice sheets clearly had insufficient time to melt significantly and therefore the presence of major ice sheets at high northern latitudes (Figure 7) did not hinder major atmospheric CH_4 increases (just as they did not during lesser interstadial CH_4 increases).

The large, rapid CH_4 increases during the late Quaternary have generally been attributed to increased emissions from tropical and high latitude wetlands in varying proportions [Petit-Maire et al., 1991; Chappellaz et al., 1990; Severinghaus et al., 1998; Severinghaus and Brook, 1999; Brook et al., 1999; 2000; Maslin and Burns, 2000]. We term this the *Wetland Methane Hypothesis*. Because ice sheets covered most of the present day northern wetlands during the last glacial episode (Figure 7), earlier studies usually invoked low latitude wetlands as the major source of CH_4. However, recent interpretations of inter-hemi-

Methane Hydrates in Quaternary Climate Change
© 2003 by the American Geophysical Union
10.1029/054SP05

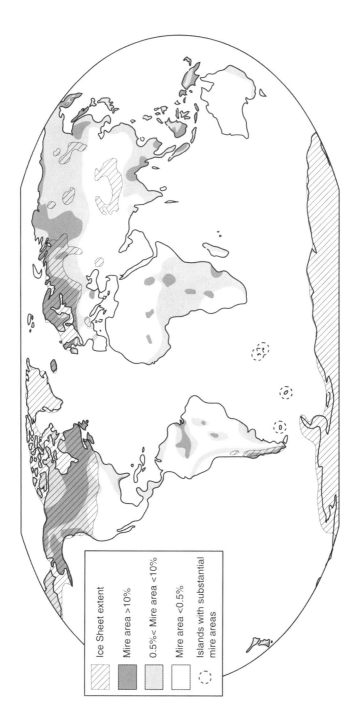

Figure 7. Global distribution of modern mires (a variety of different wetland systems) [Gore and Goodall, 1983a,b] superimposed on Last Glacial Maximum ice sheet extent [Adams et al., 1990].

spheric CH_4 gradients have suggested high to middle northern latitudes as the primary source region in spite of the vast ice sheets covering much of the Subarctic region [Dällenbach et al., 2000]. We will show the *Wetland Methane Hypothesis* to be an untenable model that is not supported by the geologic data.

Methane hydrate degassing has generally been considered as an improbable CH_4 source for the rapid rises [Chappellaz et al., 1993; Brook et al., 2000]. Blunier [2000], for instance, considered it unlikely that methane hydrate releases were regular enough to maintain an elevated atmospheric CH_4 concentration. Other potential emission sources such as termites, ruminants, natural gas seeps, or wildfires were inadequate to cause and maintain CH_4 mixing ratios of 650 ppbv [Nisbet, 1992]. At present, the ocean is considered to be only a minor source of CH_4 (5 to 20 Tg yr[-1]) [Watson et al., 1990], although this is likely highly underestimated [Hovland et al., 1993]. Quay et al. [1991] and Hovland et al. [1993] have suggested a much larger oceanic CH_4 contribution (50 to 65 Tg yr[-1]), largely from natural gas seeps.

The prevailing consensus is that the rapid CH_4 increases associated with late Quaternary glacial terminations resulted from increased wetland development (*Wetland Methane Hypothesis*) due to sudden increases in precipitation and temperature associated with a switch in climate state. This hypothesis infers that: (1) a major increase in rainfall over continental regions resulted from atmospheric warming and circulation changes, including strengthening of tropical monsoons; and (2) extensive mature wetland systems already existed and were essentially poised, during glacial and stadial terminations, to rapidly produce CH_4 at near-Holocene levels in just a few decades to centuries. In contrast, we argue from a wide range of geological evidence presented below, that these rapid CH_4 increases could not have come from continental wetlands.

If the *Wetland Methane Hypothesis* is correct and most CH_4 was produced in wetland systems throughout the entire late Quaternary, there should be a significant body of geological evidence indicating the presence of wetlands of sufficient extent to produce the rapid and major rises in CH_4 recorded in ice cores. Did this CH_4 come from wetlands? To test the *Wetland Methane Hypothesis* we conducted a detailed search of the literature and assembled the most comprehensive global synthesis of geological data yet available for late Quaternary wetland development. It was necessary to investigate the history of all large wetland areas that may have been major CH_4 sources at any time during the late Quaternary and also consider the possibilities of previous source regions that are no longer important. We also have summarized modern processes of CH_4 production in modern wetland systems to better evaluate the *Wetland Methane Hypothesis*.

Associated Precipitation Increase during Rapid Warming Episodes

Strong evidence exists that interstadials and interglacials experienced higher precipitation compared with glacials and stadials. Precipitation increased over large areas at stadial terminations and during the last deglaciation. Rapid shifts in atmospheric systems led to strengthened tropical monsoons [Xiao et al., 1999; Y. Wang et al., 2001]. It is likely that wetland ecosystem development and CH_4 production were affected by these precipitation increases although the extent of wetlands and their CH_4-production potential were controlled by a complex array of factors other than precipitation as discussed later.

Evidence for sudden, major river runoff increases from northern South America is recorded at the onset of interstadials in Cariaco Basin [Peterson et al., 2000] (Figure 8). Likewise, in marine sequences off northeast Brazil, interstadials were marked by increased fluvial input of terrigenous sediments as well as pollen and spores of moist vegetation indicating wetter conditions [Behling et al., 2000] (Figure 9). The last deglacial was an interval of especially high continental fresh-water discharge from northern South American rivers, inferred to have resulted from inferred higher continental precipitation. In this region, an increase in precipitation seems to have begun early (~16.7 ka) during deglaciation based on rapid $\delta^{18}O$ depletion in planktonic foraminifera from the Cariaco Basin interpreted by Lin et al. [1997] to reflect lower surface salinities because of significant continental freshwater runoff. Deglaciation of the northern Andes, initiated about this time [Clapperton, 1993], may have contributed to this increased runoff, and again at the end of the Younger Dryas [Maslin and Burns, 2000]. Maslin and Burns [2000] have estimated a 40% increase in Amazon River discharge at the end of the Younger Dryas. This reflects sudden northward movement of the Intertropical Convergence Zone at the onset of interstadials causing major reorganization of precipitation patterns on the continent [Peterson et al., 2000] (Figure 8). However, regional precipitation patterns did not change identically, since certain areas experienced higher precipitation and others lower [Martin et al., 1997; Thompson et al., 1995, 1998; Behling, 1997; Betancourt et al., 2000; Baker et al., 2001]. Nevertheless, the inferred changes in precipitation over broad areas of North and South America [Peterson et al., 2000] exhibit remarkable coherency with the millennial-scale climate variability exhibited in Greenland ice cores.

Increased precipitation during interstadials also resulted from sudden switches in the strength of summer monsoons associated with initiation of rapid warming in Southeast Asia [Wang et al., 1999a, b; Y. Wang et al., 2001] and North Africa [deMenocal et al., 1993]. It is clear, however, that there were regional differences in response to the warming episodes. In the Arabian Sea, geochemical studies of sediments indicate that precipitation associated with intensification of the southwest monsoon was delayed and occurred in two abrupt steps. The last of these was at the end of the Younger Dryas and was followed by an especially large increase in the early Holocene at 8.8 ^{14}C kyr B.P. [Sirocko et al., 2000]. Thus, this region does not seem to have participated in major precipitation increase during early deglaciation. In North America, there is evidence for increased precipitation during early deglaciation between ~16 and 14 ka in the southeast United States [Grimm et al., 1993; Kneller and Peteet, 1993] and

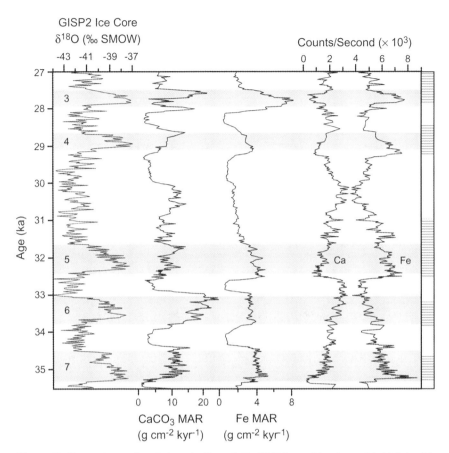

Figure 8. Comparison of variations in Fe and Ca XRF intensities (count/s) (right) with calculated mass accumulation rates (MAR) of Fe and CaCO$_3$ in Ocean Drilling Program Hole 1002C, Cariaco Basin, northern Venezuela, for the interval between 27 and 35.5 ka encompassing interstadials 3 through 7 in the GISP2 ice core δ^{18}O record (left) [after Peterson et al., 2000]. Laminated sediment intervals indicated on the right. Measured Fe intensities represent relative abundance changes in terrigenous sediments, while Ca intensities correlate with carbonate content of sediments. The accumulation of both Fe and CaCO$_3$ clearly increased during interstadial (warm) episodes. Increased terrigenous input reflects increased regional precipitation and riverine discharge and increased CaCO$_3$ accumulation reflects river-derived nutrient increases into the basin.

the Western Great Basin [Benson et al., 1992]. It therefore appears that, although changes in precipitation occurred over broad areas during the broad deglacial interval, the character and timing of these changes varied regionally and were not generally synchronous with the abrupt increase in CH$_4$ that occurred at the onset of the B-Å.

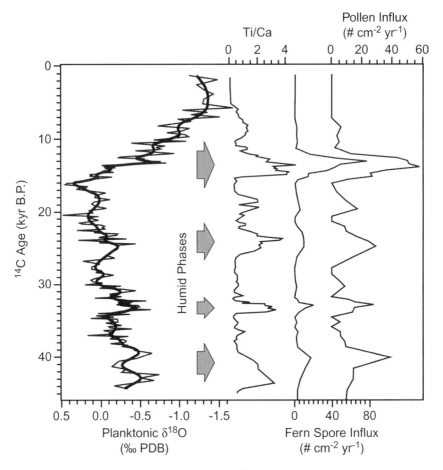

Figure 9. Record of paleoclimate (planktonic δ18O), continental sediment influx (Ti/Ca ratio), and fern spore and pollen influx into the northeast Brazilian margin for the last 45 kyr from Core GeoB 3104-1 [Behling et al., 2000]. Humid phases (arrows), indicated by relatively high Ti/Ca ratio and high influx of fern spore and pollen, reflect intervals of strong riverine discharge from northern South America during interstadials and the deglacial episode.

The evidence for rapid, millennial-scale variability in precipitation has led to a general but mistaken consensus (see below) that associated wetland development produced the rapid atmospheric CH_4 increases during the late Quaternary. The *Wetland Methane Hypothesis* assumes that wetlands were well established or rapidly expanded at glacial and stadial terminations in response to increased precipitation. Sufficient organic accumulation necessary for large-scale CH_4 production must also have occurred.

However, within this framework, it is important to consider two aspects related to increased precipitation and wetland expansion. First, high annual rainfall by itself is insufficient to promote the formation of swamp and peatland. The frequency and seasonal distribution of rainfall is much more important than the total annual volume [Thompson and Hamilton, 1983]. Rainfall frequency and reliability is critical in swamp and peatland formation. These environments develop when rainfall is spread throughout the year or, on the equator, during two annual wet seasons. Second, any increase in precipitation during the last glacial episode would not necessarily have promoted the development of extensive CH_4-producing wetlands because of efficient drainage of the continents by incised rivers when sea level was low. A major increase, followed by gradual decrease in terrigenous sediment export to the Brazilian continental margin occurred during the deglacial transition into the Holocene [Behling et al., 2000] (Figure 9). This trend appears to have resulted from initially increased continental sediment erosion and seaward transport during the marine transgression and warming, followed by increased continental storage associated with extensive wetland development and sediment accumulation in equilibrium with the higher base level. Understanding the origin and evolution of wetland systems during the late Quaternary is critical for resolving the origin of the rapid atmospheric CH_4 oscillations marking the late Quaternary.

Modern Wetland Distribution

Present day natural wetlands (Figure 7) are major contributors to the modern CH_4 budget [Cicerone and Oremland, 1988; Schütz et al., 1990]. The annual CH_4 emission from wetlands is estimated at between ~110 and 115 Tg [Matthews and Fung, 1987; Cicerone and Oremland, 1988; Prather et al., 1995, 2001; Matthews, 2000]. Other natural modern sources total only 45 Tg. Methane emissions from natural wetlands have been estimated to only be between 10 and 28% of all modern CH_4 sources [Cicerone and Oremland, 1988; Prather et al., 1995, 2001; Van den Pol-van Dasselaar and Oenema, 1999], however, because anthropogenic sources currently dominate the system [Prather et al., 1995]. The contribution from natural wetlands dominated the system before anthropogenic sources developed historically. Given this importance, we will summarize present-day wetland distribution and evidence for major changes in wetland distribution associated with late Quaternary climate evolution.

A diverse and complex set of terms are used to describe and categorize wetlands, including: mires, bogs, swamps, marshes, fens, and peatlands. Most importantly, all of the wetland systems that we refer to in the following synthesis are CH_4-producing ecosystems. Definitions may vary between disciplines and regions [Gore and Goodall 1983a, b]; as much as possible, we follow the terminology of the cited authors.

Methanogenesis, the biological production of CH_4, takes place under anoxic conditions in ecosystems that are chronically flooded and have sufficient carbon substrate and levels of microbial activity to utilize the oxygen present [Schimel et al., 1993; Valentine et al., 1994; Bohacs and Suter, 1997]. Appropriate redox potential for methanogenesis requires saturated or near-saturated soils and an absence of competing, more electronegative electron acceptors (SO_4^{2-}, NO_3^-, Fe^{3+}) [Valentine et al., 1994]. Methane production and emission is controlled by various factors dependent on wetland ecosystem structure and their microbial communities [Conrad, 1989; Cao et al., 1998]. In addition to the availability of organic substrates and electron acceptors, CH_4 production depends on appropriate temperature and pH [Conrad, 1989]. Most organic carbon substrates must be fermented to produce either acetate or H_2 before being converted to CH_4, a major process in terrestrial wetland ecosystems [Schimel et al., 1993; Valentine et al., 1994]. Methane fluxes to the atmosphere are largely controlled by soil moisture content [Roulet et al., 1992; Keller and Reiners, 1994]. In general, flooded soils can produce 100 times more CH_4 than dry soils, and warmer wet soil further increases CH_4 emissions. In contrast, the development of warmer, drier soils can lead to decreased CH_4 emissions [Cao et al., 1998].

Much CH_4 emitted from wetland soils is oxidized in the surface aerobic layer and within vascular plants [Reeburgh et al., 1993; Schimel et al., 1993, and references therein]. A high proportion (40 to 95%) is consumed by CH_4-oxidizing bacteria in the upper levels of the soil, sediment, or water column before it can be released into the atmosphere [Galchenko et al., 1989; Orphan et al., 2001]. Only a small fraction of that produced reaches the atmosphere [Fung et al., 1991; Yavitt et al., 1988]. Most aerobic soils do not support CH_4 production and emission, instead being important sites for deposition and microbial oxidation of atmospheric CH_4 [Cicerone and Oremland, 1988; Conrad, 1989]. In the Atlantic lowlands of Costa Rica, soil consumption of atmospheric CH_4 is greatest when soils are relatively dry [Keller and Reiners, 1994]. Forest soils consume CH_4, while poorly drained pasture soils generally produce CH_4. Tropical seasonality in precipitation and hence in soil moisture strongly affects CH_4 production. Drying of tropical wetlands can lead to aerobic conditions associated with drier soils.

Extensive seasonal wetland areas occur in floodplains of large rivers such as the Amazon. As a result, vast areas can seasonally switch from significant CH_4 sources to weak sinks [Rodhe, 1990; Tathy et al., 1992; Keller and Reiners, 1994]. The emission period of tropical wetlands is 180 days yr^{-1}, about the duration of the wet season [Rodhe, 1990]. Increased precipitation alone, however, does not necessarily cause increased CH_4 emissions. This requires chronic flooding of organic rich soils such as peat, illustrated by the rise in CH_4 since ~1750 which has been linked to expansion of rice cultivation; that is increased development of wetland systems [Blunier et al., 1993].

Laboratory and field studies have demonstrated that the position of the water table is a vital control on CH_4 emissions from peatlands and other wetland systems [Peterson et al., 1984; Moore and Knowles, 1989; Bubier and Moore, 1994; Whalen et al., 1996; Bohacs and Suter, 1997; Kettunen et al., 1999] and much more important than the temperature regime. Water table fluctuations influence both the thickness of the soil or peat profile in which methanogenesis is possible, and also that of any overlying aerobic zone where CH_4 oxidation occurs [Valentine et al., 1994]. Even slight lowering of the water table can lead to CH_4 oxidation and transformation of a CH_4 source to a sink. Indeed, it has been suggested that warming of wet tundra areas, in the absence of compensating higher precipitation, can cause lowering of the water table, aeration and decomposition of peat and a significant decrease in CH_4 emissions [Oechel et al. 1993; Oechel and Vourlitis, 1994; Cao et al., 1998]. Thus, past warmer periods of peat accumulation must also have been wetter. Increased temperature (>2°C) in the absence of increased precipitation will reduce CH_4 emissions because of associated reduction in soil moisture [Cao et al., 1998].

Modern wetlands represent a major ecosystem (Figure 7) currently occupying ~5.3 × 10^6 km² [Matthews and Fung, 1987] or ~15% of the total land surface [Velichko et al., 1998]. Wetland CH_4 is dominantly produced in two main regions: the northern high latitudes and the tropics-subtropics (Figure 7). The greatest aerial extent of modern wetlands is at high northern latitudes [Matthews and Fung, 1987; Aselman and Crutzen, 1989; Chanton et al., 1995]. These high-latitude wetlands are dominated by peatlands. Peatlands are wetland ecosystems where productivity exceeds biodegradation, an imbalance that causes organic matter accumulation [Williams and Crawford, 1984]. Peatlands account for 85% of Canadian wetlands [Roulet et al., 1993, 1994] as well as most wetlands north of 45°N [Matthews and Fung, 1987; Chanton et al., 1995]. The most extensive peatlands in the world occur in the boreal taiga zone between 45° and 65°N. More than 95% of the world peat resources occur in Russia, Canada, Scandinavia, Finland, the United States, and United Kingdom [Gore and Goodall, 1983a, b; Harriss et al., 1985]. Northern peatlands are associated with permafrost or perennially frozen ground [Brown, 1968].

Expansion of the northern peatlands is an interglacial phenomenon, continuing today and likely contributing to a long-term increase in atmospheric CH_4 concentrations [Harriss et al., 1985; Crill et al., 1988]. Wetland ecosystems take many forms that vary in their CH_4 emissions. Despite the importance of CH_4 emissions from wetlands, considerable uncertainties exist in the flux estimates from different sources [Cao et al., 1998]. Roulet et al. [1992] found that the most significant CH_4 emitters in central Ontario, Canada were beaver ponds followed by thicket swamps and bogs. Mixed swamps, marshes and fens emitted little CH_4 in this area.

Estimates of annual emissions of CH_4 from modern natural wetlands in different latitudes differ significantly amongst workers. Early estimates [Matthews and Fung, 1987] suggested that the greater proportion (60%) of total CH_4 emissions (63 Tg yr⁻¹ of a total of ~110 Tg yr⁻¹) comes from peat-rich bogs between 50°

and 70°N in Canada, Alaska, and Russia (Figure 7), which comprise ~50% of total wetland area. They estimated that tropical-subtropical peat-poor swamps from 20°N to 30°S, in southeast Asia, Amazonia, and Africa (~30% of the global wetland area) account for ~30% (32 Tg yr[-1]) of total wetland emissions. These estimates of the relative role of high latitudes and tropical areas are opposite to those of Bartlett and Harriss [1993], who determined a greater proportion (60%) of total CH_4 emissions (66 Tg yr[-1] of a total of ~109 Tg yr[-1]) from tropical wetlands and only ~34% (38 Tg yr[-1]) from high northern latitudes. The estimates of Bartlett and Harriss [1993] are similar to those of Cao et al. [1998], who also estimated a greater tropical wetland source (51 Tg yr[-1] of a total of 92 Tg yr[-1]), relative to high latitude northern wetlands (24 Tg yr[-1]) and temperate wetlands (17 Tg yr[-1]).

Inconsistencies of Wetland Methane Hypothesis with Rapid CH_4 Increases

It is generally assumed that tropical and high latitude wetlands, the dominant modern CH_4 sources, were also the sources for the rapid atmospheric CH_4 increases during glacial and stadial terminations. However, strong evidence exists suggesting that wetland ecosystems in these areas were not extensive enough, nor sufficiently developed, to produce the fluxes needed to sustain the rapid increases in atmospheric CH_4. We follow Nisbet [1992] who suggested that the suddenness of the atmospheric CH_4 increase at Termination 1A is difficult to explain biologically. No evidence exists to support the hypothesis that biological CH_4 sources during the B-Å were much greater than during the LGM.

The major modern high latitude CH_4-producing system (Figure 7) was not available during the last glacial episode because it was largely covered by the North American and Baltic ice sheets [Nisbet, 1992]. Furthermore, as detailed later, paleoecological evidence from areas in northern Eurasia not covered by ice during the last glacial episode suggests that these areas did not contain any major wetlands like today. As a result, this region was marked by lower CH_4 production compared with the present [Velichko et al., 1998]. Thus, in light of these observations, it is difficult to invoke a major terrestrial northern high latitude CH_4 source during interstadials of the last glacial episode, despite an inter-polar atmospheric CH_4 gradient that suggests that a northern high latitude source was present [Dällenbach et al., 2000].

Modern wetland systems in tropical-subtropical regions (Figure 7) also did not seem to be sufficiently developed during the last glacial episode to account for the sudden atmospheric CH_4 increases. Low sea levels (~120 to ~60 m) dropped base levels of erosion and caused river down-cutting, forming gorges across continental shelves and active cutting of submarine canyons on the continental margins [Kennett, 1982]. The incision of river systems led to higher river gradients and more efficient continental drainage. This would have strongly decreased flood plain extent and caused a major reduction in CH_4-producing wetlands. During glacial episodes, the Amazon River flowed in a gorge across the continental shelf [Colinvaux et al., 1996] increasing the efficiency of drainage in Amazonia. Significant infilling of fluvial networks in response to deglacial sea-level rise could not have occurred immediately and, in fact, did not occur until within the Holocene [Chen et al., 1999]. Changes resulting from sea-level rise, sediment infilling of river channels, and floodplain formation took several thousand years. Sea-level stability finally occurred during the middle and late Holocene creating conditions that were especially favorable for major wetland development.

Wetlands are complex ecosystems that require significant time for initial development and later expansion. This is supported by a wide range of evidence indicating that large modern floodplains and coastal wetlands with their associated organic carbon accumulation did not begin forming until within the Holocene. Valentine et al. [1994] have suggested that CH_4 production is an ecosystem process coupled to plant species composition, primary production, and overall heterotrophic metabolism, and also linked to climatic, hydrologic, and other factors that influence the decomposition of organic matter. Response to any improvement in the physiological environment for methanogenesis is limited by other factors such as low rate or poor quality of substrate supply, which can severely restrict fermentation [Valentine et al., 1994]. Important feedback mechanisms exist between the physical environment and wetland expansion. For example, the development of wetland vegetation in river systems helps impede river flow, which, in turn, leads to further wetland expansion. A feedback loop exists between the presence of swamps and the stability of river levels. Swamps can assist with stabilization of river level that, in turn encourages the growth of permanent swamps [Thompson and Hamilton, 1983]. Swamps act as "sponges" in river systems and therefore help regulate river flow. Thus, it is difficult to envision a process of voluminous CH_4 production from continental wetlands that was so sensitively poised to switch on in response to climate change, when so many other factors contribute to the magnitude of global wetland CH_4 production.

Critical Role of Sea-Level Rise in Wetland Formation

During the LGM, most lowland reaches contained incised valleys that had only limited wetlands. Rivers transported their sediment load to the edge of the continental shelves. Continued rejuvenation of individual drainage systems due to glacial sea-level fall was mostly focused in coastal regions, although the greatest effects occurred where base level change was large, incision rapid, and rivers confined [Schumm, 1993]. During the sea-level rise associated with the last deglaciation, increased accommodation space refocused sediment deposition on the inner shelves and expanding estuaries as freshwater and brackish tidal marsh/estuarine units. The associated change in base level elicited highly varying responses from rivers, depending on a number of factors including relief, river gradient, quantity of sediment delivered from source areas, climate, and bedrock and structural controls [Schumm, 1993]. For most alluvial rivers, the effects of a base level change can be significant, with resulting major expansion of wetland systems as flood plains developed during sea-level rise.

The distribution of peat-forming mires is strongly controlled by the position of the groundwater table. Over broad areas, including near coastal and paralic settings, the groundwater table is, in turn, strongly controlled by sea level and the precipitation/evaporation ratio [Bohacs and Suter, 1997; Hay and Leslie, 1990]. Lowered sea level rapidly draws down the coastal aquifer and lowers the groundwater table. This reduces groundwater discharge, which migrates seaward. Sea-level fall effectively restricts the occurrence and robustness of coastal-mires and decreases accommodation space, thus diminishing the likelihood of important peat accumulation [Bohacs and Suter, 1997]. In contrast, sea-level rise leads to increasing accommodation space and rising water tables. The interface between salt and freshwater migrates landward and upward. Bohacs and Suter [1997] suggested that relative sea-level rises influence accommodation space and accumulation of peat at least 40 km inland. Widespread mires become established only late in a transgression when sea-level rise decelerates. At these times, peat accumulation rates are able to keep up with slower sea-level rise. During the latest Quaternary, such conditions did not begin to develop until the early Holocene. Peat-forming ecosystems require sufficient time to become established and to accumulate organic matter. If increases in accommodation space and/or environmental change occur too rapidly, peat does not accumulate [Bohacs and Suter, 1997].

Wetlands of Modern Deltas

Major modern coastal wetland systems are associated with river deltas. These wetlands accumulate significant volumes of organic material including peat. Because of this, deltaic wetland systems such as those of the Mississippi and Yangtze Rivers can produce large volumes of CH_4 and exhibit features associated with gas escape (e.g., pockmarks, collapse depressions) [Hovland et al., 1993, and references therein]. Wetlands associated with modern river deltas are very extensive and must contribute significantly to atmospheric CH_4 levels. Nevertheless, these wetland systems also formed late, within the Holocene [Stanley and Warne, 1994; Stanley and Hait, 2000]. Development of modern deltas occurred as a result of post-glacial sea-level rise that was required to produce sufficient accommodation space for sediment deposition on the inner continental shelf. Sea-level rise was initially too rapid to allow significant accumulation of deltaic sediments. From a study of 36 modern deltas, Stanley and Warne [1994] and Stanley [2001] concluded that deltaic sediment accumulation began between 9.4 and 7.4 ka (mode = ~8 ka). Holocene deltas began to form globally at about the same time [Stanley, 2001]. Sediment accumulation began only as a result of the deceleration of sea-level rise.

Although sea level was the principal factor involved in delta formation, it is clear that other climate related factors played a role in controlling the accumulation of deltaic sediments. For example, the development of the Ganges-Brahmaputra River delta began in the early Holocene (~11 ka) when rising sea level flooded the Bengal Basin, thereby trapping most of the river discharge on the inner continental shelf [Goodbred and Kuehl, 2000]. Peak delta sediment accumulation, however, was delayed until between ~9 and 6.5 ka in the early Holocene. This timing not only coincided with sea-level deceleration [Stanley and Warne, 1994], but also was associated with a peak in precipitation and river runoff during intensification of the southwest Asian monsoon. Late Quaternary monsoon strength was centered at a 9 ka peak in regional insolation [Goodbred and Kuehl, 2000, and references therein]. This event coincided with widespread (India, Arabia, North Africa) increases in precipitation and associated fluvial incision in humid/sub-humid tropics during the early Holocene [Sirocko et al., 1993; Goodbred and Kuehl, 2000].

The Fly River Basin, New Guinea is a well-documented example of late Quaternary sedimentation under strong control by sea-level rise [Harris et al., 1996]. During the last glacial episode, the river was well incised with fluvial incision probably most extreme in its lower reaches particularly when sea level was more than 50 m below modern levels. Sea-level rise during deglaciation caused sediment deposition well inland in the incised river valleys, starving the continental shelf during the Holocene. Although sediment accumulated in the Fly River Basin during the latest Quaternary, major wetland formation was limited to the Holocene [J. Chappell, personal communication].

Although development of wetland and deltaic systems are closely related, CH_4-producing wetlands in deltas require time to form. In the Mississippi River Delta, significant peat did not accumulate until the late Holocene [Kosters and Suter, 1993] when favorable hydrological conditions developed. Critical conditions arose as fresh groundwater was recharged into the delta plain sediments, levees formed, and standing bodies of water developed to allow the anoxia necessary for organic buildup. Thus, it appears that the modern deltaic wetland systems did not begin to form during the deglacial sea-level rise, but were delayed until after sea-level rise began to decelerate and stabilize in the late early to middle Holocene.

Limited Wetlands on Exposed Continental Shelves

Continental shelf areas, formerly exposed during the low sea levels of the last glacial episode, are possible candidates for earlier (pre-Holocene) wetland development. Some large areas in the tropics, such as the Sunda (Figure 10) and Sabul Shelves of Indonesia, and high latitudes such as the Bering Strait (Beringia) in the

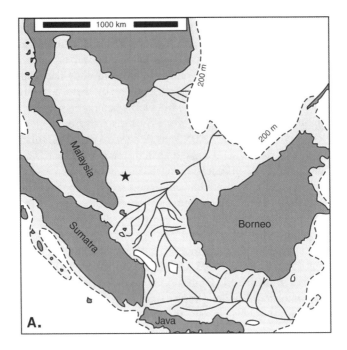

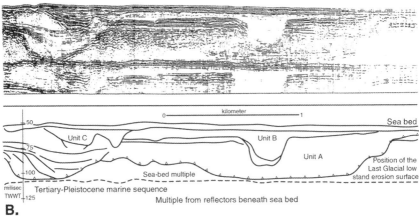

Figure 10. (A) Extent of subaerially exposed continental shelves (note particularly the vast Sunda Shelf) around Indonesia, Malaysia and Vietnam, during the Last Glacial Maximum sea-level lowstand showing the major river systems that cut these shelves [after Stattegger et al., 1997, and references therein]. (B) Seismic profile across infilled Pleistocene channels in the Sunda Shelf east of the Malaysian Peninsula (general location indicated by star) [Evans et al., 1995; Stattegger et al., 1997]). Incised valley systems up to 50 m deep were cut during the low sea-level stand of the Last Glacial Maximum.

Northeast Pacific, potentially could have been areas of extensive wetland development, now inundated as a result of deglacial sea-level rise. Determining wetland development and extent in shelfal areas subjected to marine transgression is difficult because scouring may have removed earlier evidence of wetland development [Elias et al., 1996]. Nevertheless, paleoenvironmental conditions of these shelves have become sufficiently well known to at least determine the relative extent of wetlands prior to transgression and it appears that no major peatland systems occurred on continental shelf areas during the LGM and glacial termination. Peat layers that are preserved are thin and patchy suggesting that the environment was not conducive for producing significant wetland systems.

During low sea-level stands of the last glacial episode, very broad areas of the continental shelf were exposed in Beringia, including the north Bering Shelf, the Bering Strait Shelf, and the Chukchi Shelf, an extensive area of potential wetland development. Full glacial conditions were marked by relatively dry tundra heaths or dry meadows, interspersed with marshes and small ponds. Climate conditions during deglacial times (~15 to 11 ka) were significantly warmer and moist. Evidence from sediment cores indicates that peat accumulation did occur in the region [Elias et al., 1992, 1996] during the last glacial and deglacial episodes. Most peat has been dated to between 14 and 9 ka. However, the peat layers are thin (<10 cm), and laterally discontinuous throughout the region [Nelson, 1982; Nelson & Nio, 1982; Elias et al., 1992, 1996]. Small-scale patchy wetlands are suggested, rather than extensive wetlands important for significant CH_4 production.

The Sunda and Sabul Shelves most likely did not harbor extensive wetland areas during the last glacial episode because of the development of well-incised drainage networks [Verstappen, 1975; Tjia, 1980; Hanebuth and Stattegger, in press] up to 50 m deep [Stattegger et al., 1997] (Figure 10). These and other shelf areas were, as a consequence, well drained, with minimal wetland development. During the lowstand of the last glacial episode, the shelf was marked by widespread soil formation, including lateritic soils, with no evidence of significant peat accumulation or organic-rich layers [Hanebuth and Stattegger, in press]. It appears that low-stand vegetation was relatively dry and precipitation was lower than during the Holocene. The shelf was a mosaic of extended savanna (dense grassland with bushes) and local patches of tropical rainforest or dry forest [Hanebuth, 2000]. Extensive sediment bypassing of the shelf to the South China Sea occurred.

The interval of deglacial sea-level rise on the Sunda and Sabul Shelves was marked by many changes, including significant vertical and lateral facies changes, and extreme pulses and/or hiatuses in sediment deposition. The rapidly changing conditions were not conducive to important peat formation [Hanebuth et al., 2000]. Evidence exists in some submarine cores for the presence of mangrove swamps (which are not CH_4-producers [Harriss et al., 1988; Crill et al., 1991]) and mud flats on the Sunda Shelf associated with the transgression of the last deglaciation [Hanebuth et al., 2000; Sun and Li, 1999]. Wetlands on this shelfal area were apparently too limited during the last glacial and deglacial episodes to have had any significant effect on atmospheric CH_4 levels. Furthermore, the extremely rapid sea-level rise associated with the onset of the B-Å (16 m in 300 yr from 14.6 to 14.3 ka) [Hanebuth et al., 2000] would have been too rapid to foster organic buildup necessary for major CH_4 production. During the Holocene, the shelf became starved of sediment as wetlands formed on coastal plains.

A diversity of geological evidence also suggests that extensive wetlands did not form on the exposed continental shelves at mid-latitudes off the eastern United States during the glacial sea-level lowstand. Peat collected at numerous locations on the now submerged shelf [Emery et al., 1967] has a maximum age of 12 ka and suggests thin and patchy distribution. Rivers that crossed the shelves during the last glacial episode cut well-incised river valleys, which are suggestive of well-drained conditions. For example, the Delaware River channel was 10 to 15 m deep at mid-shelf and 30 m deep across the outer shelf [Twichell et al., 1977].

Significance of Ice Age Dryness

Much geological evidence clearly indicates that the LGM was significantly drier over broad regions of the Earth [Sarnthein and Diester-Haass, 1977; Adams et al., 1990; Crowley, 1995; Petit-Maire, 1999]. Desert and savanna environments were greatly expanded (Plate 2). Drier vegetation of deserts and arid scrub were much more extensive than during the Holocene [Adams et al., 1990; Petit-Maire, 1999]. The major modern vegetation belts did not simply move closer to the tropics when ice sheets covered the northern high latitudes. Instead, major vegetation changes occurred because of the significantly drier conditions during the LGM. Thus, polar deserts expanded in northern Asia, and mid-latitude Northern Hemisphere deserts expanded in North America, China and Africa (Plate 2). The extent of wooded steppes also increased, as did savannas [Adams et al., 1990; Petit-Maire, 1999]. The America's were relatively cold, dry and windy over broad areas of the middle and high latitudes during the last glacial episode [Muhs and Zárate, 2001]. Much of Europe was drier south of the ice sheets, such as in central France, which was marked by both low temperature and arid climate [Fauquette et al., 1999]. Extreme dryness led to widespread deposition of loess and the atmosphere became extremely dusty [Taylor et al., 1993]. Major calcium increases (transported as $CaCO_3$) recorded in Greenland ice cores during the LGM reflect increased dust transport to the region [Alley, 2000].

Plate 2. Comparison of the distribution of principal vegetation types and ice sheet extent between (A) modern and (B) Last Glacial Maximum (LGM) [after Adams et al., 1990]. The LGM was marked not only by greater ice cover, but also by much wider distribution of dry vegetation biotopes, including those of deserts, grasslands and savanna, semi-desert and dry steppe, polar desert and southern steppe tundra compared with the present day. During the LGM, the earth was largely dominated by dry vegetation biotopes.

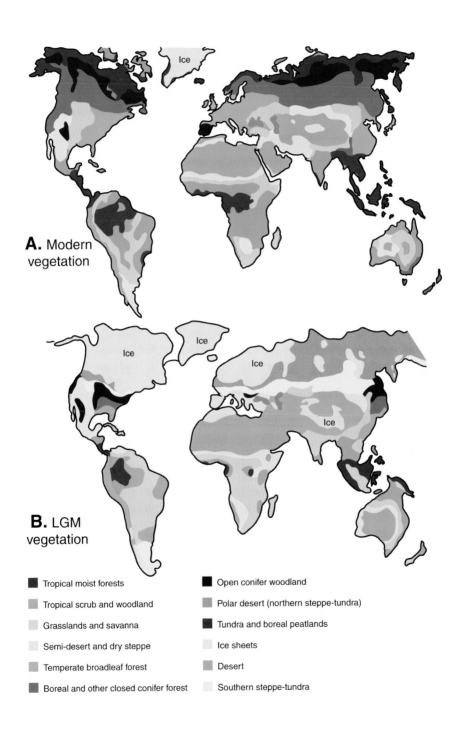

A. Modern vegetation

B. LGM vegetation

■ Tropical moist forests

□ Tropical scrub and woodland

□ Grasslands and savanna

□ Semi-desert and dry steppe

□ Temperate broadleaf forest

■ Boreal and other closed conifer forest

■ Open conifer woodland

□ Polar desert (northern steppe-tundra)

■ Tundra and boreal peatlands

□ Ice sheets

□ Desert

□ Southern steppe-tundra

A reduction in African and Asian monsoon strength during the LGM, led to an increase in aridity over broad areas [Petit-Maire, 1986, 1999]. Even in the region of the modern "warm-pool" of southeast Asia, the exposed continental shelves appear to have had relatively dry vegetation [Hanebuth and Stattegger, in press]. Pollen evidence from the South China Sea indicates that the exposed northern continental shelf was covered mainly by grasslands and adjacent elevated areas were dominated by montane conifer forests suggesting a cool, dry climate [Sun and Li, 1999; Sun and Luo, 2001]. This contrasts with the humid tropical forests of the interglacials, like those of the present. Evidence for dryness of the vegetation in this region during the last glacial episode is supported by an association with relatively abundant charcoal in South China Sea sediment sequences suggesting frequent occurrence of natural fires. In contrast, interglacial sediments contain little charcoal [Luo et al., 2001]. At the beginning of the Bølling (~14 ka), the dry shelf vegetation began to be replaced by mangrove swamps in association with marine transgression [Sun and Li, 1999].

In South America, much geological evidence also supports extensive dryness over broad areas during the LGM [Leyden, 1985, and references therein; Schubert, 1988]. Pollen evidence from Lake Valencia cores [Bradbury et al., 1981; Leyden, 1985] indicates glacial aridity associated with more extensive savanna over the Venezuelan Llanos [Schubert, 1988]. This aridity continued during the deglaciation with variably moist (mesic) conditions developing only by the earliest Holocene [Bradbury et al., 1981; Leyden, 1985; Schubert, 1988]. Conditions in the northern tropical Andes (Bogota Basin, Columbia) were both cooler and drier during the LGM [Mora and Pratt, 2001], but not further south in Bolivia [Baker et al., 2001]. Based on evidence from changes in lake levels of Lake Titicaca, maximum precipitation occurred at the end of the last glacial episode (25 to 15 ka) compared with lowest levels during the early to middle Holocene (8 to 5.5 ka) [Baker et al., 2001]. In Amazonia, clear evidence exists for somewhat drier conditions during the LGM [Colinvaux et al., 2000; Rühlemann et al., 2001; Behling and Hoogheimstra, 2001], although little consensus seems to exist as to how dry. Colinvaux et al. [2000] dispute the hypothesis [e.g., Williams, 1925; Clapperton, 1993] that dry vegetation (savanna) replaced closed forest during the LGM and suggest that the region was not a source for atmospheric dust. However, Behling and Hoogheimstra [2001] suggest that the Amazonian rain forest must have been reduced at this time.

Southern South America, including much of Argentina, was especially dry during the LGM [Alcalde and Kulemeyer, 1999; Carignano, 1999; Prado and Alberdi, 1999; Kröhling and Iriondo, 1999]. Eolian sand and loess deposition were widespread. Again, it was not until the early Holocene that humidity significantly increased. The deglacial episode remained relatively dry [Carignano, 1999]. Although evidence exists for subhumid climate during the B-Å [Kröhling and Iriondo, 1999], few wetlands seemed to have formed in southern South America.

Major feedbacks reinforcing cooling would have resulted from expansion of the deserts and grasslands and contraction of forests. This would have increased albedo over broad regions and reduced atmospheric water vapor by decreasing evapotranspiration [Harris and Mix, 1999]. The dry conditions of the LGM were clearly not favorable for the development of widespread wetland systems and peat accumulation. Indeed, global vegetation maps for the LGM indicate that swampy areas were extremely reduced and peat formation negligible [Adams et al., 1990; Crowley, 1995; Petit-Maire, 1999] (Plate 2). Adams et al. [1990] recorded no mapable areas of peat (at a global scale) during the LGM and suggested that extensive peat accumulation began only after the end of the glacial episode (after ~10 ka). This later development was linked with a more than doubling of organic carbon in vegetation and soils from the LGM to pre-industrial times (increase of ~280 Pg C (1 Pg = 10^{15} g) [Adams et al., 1990], increase of 530 to 1160 Pg C [Crowley, 1995]). This conclusion differs drastically from the results of modeling experiments, suggesting similar volumes of vegetation and soil carbon during the LGM and the pre-industrial Holocene [Prentice and Fung, 1990]. Paleo-vegetation mapping surely indicates that the modeled estimate must be incorrect, and that aridity strongly reduced carbon storage during the LGM.

Nevertheless, even in the absence of geological evidence for wetlands, Crowley [1995] suggested the presence of some wetlands to account for the CH_4 record from ice cores, which was about half the pre-industrial value of the late Holocene. Crowley [1995] suggested a peat accumulation of 120 Pg. However, the geological evidence strongly indicates that wetlands were highly limited during the LGM and Adams et al. [1990] suggested no significant accumulation of peat. If so, it follows that a higher proportion of atmospheric CH_4 during the LGM, compared with the late Holocene, must have been derived from sources other than continental wetlands. For example, Luyendyk et al. [2002] estimated a near doubling (from 18-48 Tg to 40-100 Tg yr^{-1}) of CH_4 sources from continental shelves exposed by sea-level fall during the LGM. Methane from exposed shelves is delivered directly into the atmosphere rather than being dissolved and oxidized in the ocean.

Age of Tropical Wetlands

Palynological evidence and radiocarbon dating of basal peat formation have constrained the timing of tropical wetland development. Little evidence exists for the existence of major late Quaternary tropical wetland systems before the Holocene. The formation of extensive modern tropical wetlands was limited to the Holocene, with most forming after the middle Holocene (Tables 1-3). No evidence exists to support the hypothesis that tropical wetlands formed or were reactivated during the last glacial termination. Furthermore, interstadials do not seem to have been associated with significant expansion of tropical wetlands because of their brevity and because environmental conditions were not conducive for extensive development [Behling et al., 2000].

TABLE 1. Timing of Wetland Development in Tropical South America.
Major wetlands expansion occurred during the Holocene, especially in the middle to late Holocene.

Area	Age (ka)	Change	Reference
Central Brazil	<6	Flood plains and associated wetlands.	Turcq et al., 1997
	<7	Swamp and marsh vegetation.	Salgado-Labouriau et al., 1998
Rio Negro River	~4	White Water to Black Water transition.	Latrubesse & Franzinelli, 1998
Upper Parana River, Brazil	8 - 5	Wide Flood plains formed.	Stevauz & Dos Santos, 1998
Northern Brazil	7 - 6	Low energy river system.	Behling & Lima Da Costa, 2000
	~ 2.5	Blocked river system; inundated forests.	Behling & Lima Da Costa, 2000
Southern and Central Brazil	7 - 4	Swamp vegetation.	Ledru et al., 1998a,b
Western Amazonia	4.7	Poorly drained wetland development.	Behling, et al., 1999; Behling & Hooghiemstra, 1999
Central Columbia Amazon	~3.3	Wetland vegetation; major seasonal flooding.	Behling et al., 1999
Eastern Amazon Basin	~ 8	Active to passive river system.	Behling et al., 2000
	~ 3	Blocked river system.	Behling et al., 2000
Columbian Llanos wetlands	<3.3	Wetland development.	Behling & Hooghiemstra, 1999
Guayana Shield	~8	Initial peat accumulation.	Schubert, 1988
Bolivian Amazonia	3	Savanna to rainforest.	Mayle et al., 2000

TABLE 2. Timing of Wetland Development in Tropical Southeast Asia and Asia.
Major wetland expansion occurred during the Holocene, especially in the middle to late Holocene.

Area	Age (ka)	Change	Reference
Indonesia: Wide-scale regional	<3.5	Peat accumulation.	Maas, 1996
	<4.5	Initiation of lowland peat swamps.	Anderson, 1983
Sarawak/Kalimantan: Basins & Coasts	<4.5	Peat accumulation.	Rieley et al., 1995
Sarawak/Kalimantan: Highlands	<9.0	Peat accumulation.	Rieley et al., 1995
East Sumatra	<6.0	Initiation of coastal swamps.	Diemont & Supardi, 1987; R. Wúst, pers. communication.
Malaysia	<~6	Peat accumulation.	R. Wúst, pers. communication
Thailand: Wide-scale regional	<6.9	Peat accumulation.	Maas, 1996
Thailand: Songhla Lake Basin	13.3	Peat accumulation.	Maas, 1996
Papua/New Guinea: Fly River	10	Wetland initiation.	Chappell et al., 1993
Papua/New Guinea: Sepik River	3.3	Wetland initiation.	Chappell et al., 1993
Vietnam: Mekong Delta	<4	Wetland development.	Maas, 1996
China: Yangtze River Delta	11	Wetland initiation.	Liu et al., 1992
	8 - 4	Maximum wetland development.	Liu et al., 1992

TABLE 3. Timing of Major Wetland Development in Africa. Wetland development was mostly confined to the early Holocene in North Africa. In the highlands of equatorial east Africa, most peat accumulation occurred during the Holocene.

Area	Age (ka)	Change	Reference
North Africa	~14	Lakes: initial rise.	Thevenon et al., 1999
	~12.8 - 8.7	Lakes: peak levels.	Kutzbach & Street-Perrott, 1985
	8.5 - 6.5	Wet episode in modern north African deserts.	Petit-Maire et al., 1991
	9.5 - 4.5		Lézine, 1989
	9 - 4	Nubian Desert.	Pachur et al., 1990
	6 - 3	Somalia.	Voigt et al., 1990
	~6.5 - 5.5	Central Sahara lake & marsh deposits.	Petit-Maire et al., 1990
Sudan, White Nile region	~9.5	Initiation of peak floods; Sudd swamps form.	Williams et al., 2000
Equatorial East Africa	<30	Dry, conifer forest; limited peat: glacial time.	Thompson & Hamilton, 1983
	~20	Cool, dry, minimum forest development, limited and slow peat accumulation, lakes dry: glacial maximum.	Bonnefille & Chalié, 2000
	16- 10	Minimal peat accumulation during instability: deglacial episode.	Bonnefille & Chalié, 2000
	<10	High peat accumulation: Holocene.	Bonnefille & Chalié, 2000

The wetland evolution of the Amazon Basin (5×10^6 km^2) has been well documented. Modern wetland regions in Amazonia are extensive, although they only represent ~5% of the Amazonian region [Devol et al., 1990]. Instead, terra firma areas are completely dominant, occupying about 95% of the Amazon Basin. The Amazonian wetland system was produced through seasonal flooding of the river flood plains, a process that extends far inland because of low topographic gradients. The flood plains have been estimated to cover 10^5 km^2 and there are at least 10^5 km^2 of lakes and swamps [Mitsch and Gosselink, 1993; Maslin and Burns, 2000]. Pronounced wet and dry seasons in Amazonia are driven by shifts in the position of the Intertropical Convergence Zone. Large annual water level fluctuations (~10 m) cause extensive inundation of large flood plains and the formation of seasonal wetlands exceeding 5 m in depth. There is little true swamp in Amazonia [Thompson and Hamilton, 1983]. During seasonal high water, the flood plains are dominated by aquatic and marsh vegetation, with extensive floating mats or meadows. The meadow plants have dense mats that are frequently anaerobic and produce CH_4 [Devol et al., 1990]. In contrast, during low water levels, the flood plains are dominated by terrestrial vegetation biotopes [Junk, 1983]. In these wetlands, peat is not important because organic accumulation is inhibited by increased organic oxidation during dry seasons that in turn is further enhanced by high tropical temperatures [Junk, 1983]. Permanent swamps and peat accumulation occur only in lakes that do not participate in the high water level fluctuations [Junk, 1983]. Large seasonal changes occur in CH_4 production in response to the annual flooding cycle [Bartlett et al., 1990]. Devol et al. [1990] determined that overall CH_4 emissions were ~2/3 lower during the low water (dry) season than during the high water (wet) season. This difference (68 versus 184 mg CH_4 m^{-2} day^{-1}) is caused by higher emissions from floating macrophyte environments [Devol et al. 1990].

In the central Brazilian area of Amazonia, swamp and marsh vegetation began to slowly accumulate at ~7 ka. It was only after 5 ka that the total area of savannas and wet soil vegetation began to reach present day levels [Salgado-Labouriau et al., 1998]. In the central Columbian Amazon (Caquetá River Basin) in western Amazonia, far from the ocean, poorly drained wetlands did not form until after 4.7 ka in the middle Holocene [Behling et al., 1999; Behling and Hoogheimstra, 1999]. Seasonal flooding of wetlands there did not develop until the late Holocene after ~3.3 ka [Behling et al., 1999]. Conditions were well drained before this. Furthermore, in southern and central Brazil, swamp vegetation commenced at ~7 ka, followed, during the next 3 kyr, by the development of semi-deciduous forest and then by galley forest [Ledru et al., 1998a, b]. Modern vegetation was established by 4 ka. In the Llanos region of Columbia, evidence exists for increasing wetness after 3.3 ka in the late Holocene [Behling and Hooghiemstra, 1999]. In the Guayana Shield, thin peat layers began forming only after ~8 ka [Schubert, 1988]. Schubert [1988] suggested that before the early Holocene the climate of the Guayana Shield did not favor peat formation because of late Pleistocene to early Holocene aridity.

A range of geological evidence from Amazonia strongly suggests delayed major floodplain development until after the early Holocene (Table 1). Junk [1983] and Turcq et al. [1999a, b] have argued that low sea-level stands associated with the LGM led to valley incisions and effective drainage of most of the contemporary wetlands of Amazonia (Figure 11; Plate 3). Lake levels were low and few wetlands existed [Turcq, 1999b]. During the LGM when sea level was 120 m below present levels, the increased gradient caused the Amazon and its tributaries to incise tens of meters below their flood plains [Betancourt, 2000] (Plate 3). Even in the earliest Holocene, sea level was still 25 m below modern levels. The incised valleys back-filled with sediment at a slower rate than the water level rise. This produced an open tributary system with lakes [Betancourt, 2000]. It was only during the late Holocene that sediment infilling of the previously incised tributary system was sufficiently advanced to constrict river flow and cause major over-bank flooding of the flood plains (Figure 11; Plate 3). Thus, the maximum extent of CH_4-producing wetlands in the Amazon Basin have depended more on sedimentological and environmental adjustment to sea-level rise than on increasing rainfall [Betancourt, 2000].

Sedimentological evidence indicates an evolution of the Tamanduá River system of central Brazil during the latest Quaternary [Turcq et al., 1997]. A dry interval from 17 to 10 ka was marked by a low water table and very rare organic-rich layers. This was followed by a wet interval from 10 to 6 ka marked by high river discharge and erosion. Although there was widespread development of Brazilian forests at that time, floodplains and associated wetlands did not form extensively until after ~6 ka (Figure 11). Sedimentation of river channels led to the formation of a stable, restricted water channel, a high water table and organic-rich deposition during inferred drier conditions of the late Holocene [Turcq et al., 1997].

A similar evolution has been inferred for the upper Rio Negro Basin [Latrubesse and Franzinell, 1998] including an earlier "white water" phase from 14 to 4 ka followed by a "black water" phase no older than 4 ka. The earlier white waters of the upper Rio Negro Basin transported abundant sediment load. The black water phase represents present day fluvial dynamics with little suspended load. These are the conditions that form floodplains and extensive wetlands.

Likewise, in the eastern Amazonia rain forest region of Caxiuanã (northern Brazil), the river systems exhibit evidence of decreasing energy through the Holocene [Behling and Lima da Costa, 2000]. The Rio Curuá River changed from an active to a passive system at ~8 ka (Plate 3). Pollen evidence indicates the transition from a well-drained (unflooded upland) rain forest with limited development of inundated forests to a low energy river at ~7 to 6 ka. Change to a blocked river with high water levels and marked increase of inundated forests did not occur until ~2.5 ka (Figure 11). This Holocene evolution of the wetland system is interpreted to have resulted from sea-level rise by Behling and Lima da Costa [2000]. The wide floodplain of the upper Parana River, Brazil, did not

A. Last Glacial (>15 Ka)

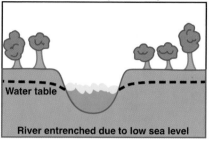

B. Mid-late Holocene (<6 Ka)

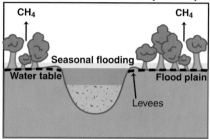

Last Glacial Hydrology

- Well drained rivers
- Incised river channels
- Dynamic flow (white water)
- Heavy sediment load
- Low water table
- No flooding of plains
- Water and sediment flushed into ocean

Mid-Late Holocene Hydrology

- Poorly-drained rivers
- Sedimented river valleys
- Sluggish flow (black water)
- Light sediment load
- High water table
- Seasonal flooding (plains innundated)
- Water entrainment and sediment trapping on plains

Last Glacial Vegetation

- Negligible wetland vegetation
- Low CH_4 production

Mid-Late Holocene Vegetation

- Extensive wetland vegetation (várzea and igapó— forests
- High seasonal CH_4 production

Plate 3. Schematic diagram of the Amazon River system contrasting hydrological conditions during the last glacial episode (>15 ka) with those during the middle to late Holocene (<6 ka). The last glacial episode was marked by incised, well-drained rivers with insignificant CH_4-producing wetland systems while the middle Holocene to present was marked by poorly-drained rivers with extensive seasonal flooding of associated flood plains leading to high CH_4 production.

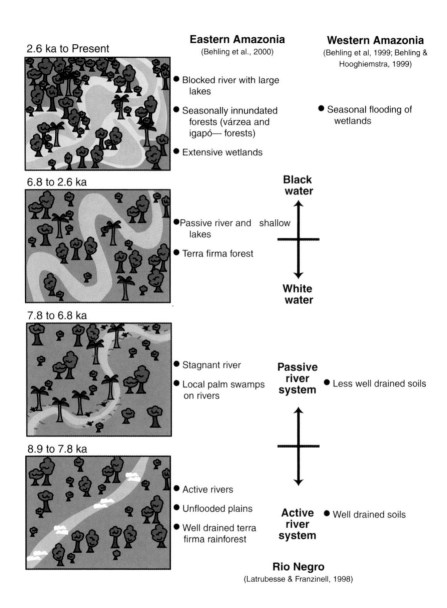

Figure 11. Schematic diagram depicting the evolution of the Amazon River system (eastern and western Amazon and Rio Negro River) from unflooded plains during the early Holocene to constricted rivers with seasonal flood plains during the late Holocene. Significant CH_4 production is inferred to have commenced at the onset of seasonal river flooding.

form until after 8 to 5 ka, at which time fluvial subtropical forest began to develop [Stevaux and Dos Santos, 1998] (Figure 11). Before this time, the river ran in braided channels.

A diversity of evidence therefore suggests that relatively active, well-drained Amazon river systems remained well into the Holocene. In spite of a general increase in precipitation during the last glacial termination, Amazonia remained well flushed, with no evidence of extensive wetland development during interstadials or Termination 1A. This is not surprising, because low sea levels and associated low base level of erosion maintained incised river channels that prevented over-bank flooding (Plate 3).

The largest wetland system of South America is the Pantanal of southern Brazil. The Pantanal is a major source region of the Paraguay River, which flows south through Paraguay and Argentina. It is a lowland area controlled by local tectonic base level, probably since the early Quaternary [M. Assine, personal communication]. Little data yet exist about the age of wetland initiation during the late Quaternary [M. Assine, personal communication]. However, Tricart [1982] and Clapperton [1993] considered the region to be relatively arid during the last glacial episode. Indeed, Tricart [1982] interpreted a mosaic of ponds in some areas to be relict eolian deflation hollows from the last glacial episode. Furthermore, central and southern Brazil, in general, appear to have been relatively dry during the last glacial episode followed by increased humidity during the early Holocene [Ferraz-Vicentini and Salgado-Labouriau, 1996; Stevaux, 1994]. However, recent evidence of regional changes in Holocene precipitation (although in a different watershed) at about the same latitude (15°S) as Pantanal comes from Bolivian Amazonia [Mayle et al., 2000], near the boundary between the Amazonia rainforest and savanna. Late Quaternary pollen evidence suggests early-middle Holocene drier conditions marked by dominance of savanna. Rainforest communities have expanded into the region only since 3 ka [Mayle et al., 2000]. The rainforest expansion was linked to increased regional precipitation at ~3 ka, influenced by southward movement of the Intertropical Convergence Zone [Mayle et al., 2000]. This correlates with evidence for an increase in snow accumulation on Sajama Mountain [Thompson et al., 1998]. Mayle et al. [2000] attribute the regional dryness during the early and middle Holocene to a minimum in Southern Hemisphere insolation between ~12 and 9 ka. This was followed by increasing insolation during the late Holocene, with resulting higher humidity and rainforest expansion.

In southeast Asia, modern peatland areas are very extensive. In Indonesia alone, peatlands are estimated to be ~17 to 27×10^6 ha^2 [Maltby et al., 1996]. Much of this wetland system, and that elsewhere in Asia, developed during the Holocene (Table 2). The development of vast alluvial deposits in river valleys, deltas, and within sheltered bays along the coast created the conditions necessary for formation of peat swamps. Lowland peat swamps, which dominate, developed during the Holocene [Anderson, 1983]. The initiation of modern peat deposits over wide areas of Indonesia was considered by Maas [1996] to be younger than ~3.5 ka. In Sarawak, Central Kalimantan and West Kalimantan, coastal peatland began to accumulate at ~4.5 ka, while the age of peat deposits at higher elevations extends to over 9 ka [Rieley et al., 1995]. Highland peat appears to have undergone most rapid accumulation between 9 and 6 ka, and have decreased since [Rieley et al., 1995]. The vast coastal swamps of eastern Sumatra have been dated to <6 ka [Diemont and Supardi, 1987; R. Wúst, personal communication]. Dated wetlands in the Mekong Delta of Vietnam are younger than 4 ka [Maas, 1996]. In Thailand, ages of peat formations are younger than 6.9 ka, although at one location (Songhla Lake Basin) peat has been dated at 13.3 ka [Maas, 1996]. Modern peat deposits in Malaysia are generally younger than the middle Holocene (<~6 ka [R. Wúst, personal communication]). Major wetland systems and underlying organic-rich sediments and peat in Papua, New Guinea also largely developed during the Holocene. Initiation of the vast wetlands of the Fly River occurred after 10 ka [Chappell, 1993], while those of the Sepik River are younger than 3.3 ka [Chappell, 1993]. Wetland development in the Yangtze River delta seems to have been somewhat earlier during the Holocene; ~11 ka, with maximum development between 8 and 4 ka [Liu et al., 1992].

The Holocene development of these wetland areas in southeast Asia has been linked to sea-level rise [Maas, 1996]. This caused a rise of the water table in coastal areas, and in turn assisted with the formation of marshy and swamp-like alluvial plains. These steps initiated the formation of future peat deposits [Polak, 1975]. Nevertheless, as elsewhere, there appears to have been a delay in accu-mulation of organic-rich deposits [Anderson, 1983], probably reflecting time required for ecosystem development.

Likewise in Africa, much of the wetland systems developed during the Holocene (Table 3). In northern Africa, lake levels began to rise at ~14 ka, in association with the B-Å warming and inferred higher precipitation [Thevenon et al., 1999]. However, lake levels did not peak until later, between ~12.8 and 8.7 ka, associated with strengthening of the summer monsoon [Kutzbach and Street-Perrott, 1985; Street-Perrott, 1992; Petit-Maire et al., 1991]. Broad areas of north Africa, now represented by the Saharan, Nubian, and Arabian deserts, exhibit evidence of relatively high seasonal rainfall from the early to middle Holocene [Petit-Maire, 1986; Petit-Maire and Riser, 1981; Lézine, 1989; Lézine and Hooghiemstra, 1990; Barker et al., 2001]. This wet episode (the African humid period) has been variously dated between 9.5 to 4.5 ka [Lézine, 1989] or 8.5 to 6.5 ka [Petit-Maire et al., 1990]. The pluvial phase in the Nubian Desert has been dated at 9 to 4 ka [Pachur et al., 1990] and for the Horn of Africa (Somalia) between 6 and 3 ka [Voigt et al., 1990]. This pluvial episode does not seem to have occurred during the deglacial interval. Much evidence exists for surface water runoff and the emergence of aquifers that resulted in small freshwater lakes. Lake or marsh deposits date to ~6.5 to 5.5 ka [Petit-Maire et al., 1990]. Little organic-rich sediment was deposited in these former lakes, probably

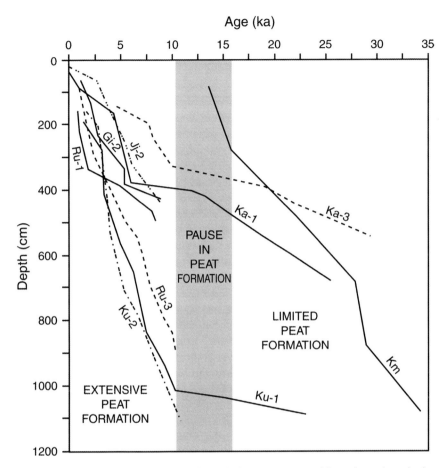

Figure 12. Ages and deposition rates of peat in 9 sequences cored from 6 peat bogs in the central East African Highlands (2° to 4°S; 1800 to 2240 m above sea level) [Bonnefille and Chalié, 2000]. Peat formation was limited and slow during the last glacial episode and highly restricted during the climate instability that occurred during deglacial times (~15 to 10 ka). Peat deposition became much more rapid and widespread during the Holocene.

because of unsuitable conditions for preservation, instability, and insufficient time for development [Petit-Maire, 1986; Schulz and Whitney, 1986]. These wetlands appear to have been unimportant for CH$_4$ production.

Swamp development in east Africa expanded in association with wet periods [Nisbet, 1992]. Peat layers in equatorial mountains of East Africa are mostly younger than 10 ka (Figure 12) [Bonnefille and Chalié, 2000] (Table 3), and expanded during peak lake levels [Street and Grove, 1979] associated with maximum monsoon strength [Short and Mengel, 1986]. Rates of accumulation of peat are significantly higher during the Holocene than during the last glacial episode (Figure 12) [Bonnefille and Chalié, 2000]. The deglacial interval, which includes the most rapid rise in atmospheric CH_4 levels (Figures 1 & 2), exhibited the lowest rates of peat accumulation (Figure 12). This was a time of large climate oscillations in lake records. Bonnefille and Chalié [2000] point out that alternating dry and wet climate conditions did not favor formation of peat during the transition. The deposition of peat requires specific stable conditions of temperature and humidity. There appears to have been general cessation or reduction in peat accumulation and distribution during the glacial termination [Thompson and Hamilton, 1983].

Africa is dry with >95% of the continent receiving <500 mm yr[-1] of precipitation. Nevertheless, wetlands of various kinds are common and equatorial Africa supports several of the world's largest swamps [Thompson and Hamilton, 1983]. Low and medium altitude grass, reed and sedge swamps are widespread. These wetlands are unimportant as producers of peat or even organic-rich soils. Peat largely accumulates at higher altitudes in central Africa and Malawi and near sea level in South Africa, where the climate is no longer of the lowland tropical type [Thompson and Hamilton, 1983]. The history of peat formation in east Africa illustrates the significant delay that can exist between increased precipitation (lake rise) and organic carbon accumulation required to produce CH_4. Wetlands in the White Nile region, Sudan, also exhibit major expansion during the early Holocene [Williams et al., 2000]. The early Holocene was wetter than any other interval of the late Quaternary. Peak floods of the White Nile began at ~9.5 ka. The vast Sudd Swamps of southern Sudan did not develop until the early Holocene and have remained even after arid climates returned at ~4.5 ka [Williams et al., 2000]. During the LGM, both Uganda and Ethiopia were drier and cooler and dry season flow in the Blue and White Nile Rivers was low or absent. Between 18 and 12.5 ka, the lower White Nile was strongly seasonal until Lake Victoria began to overflow at ~12.5 ka. Prior to this, Lake Victoria was a closed basin with no outlet to the White Nile [Williams et al., 2000, and references therein].

Age of Northern High Latitude Wetlands

Wetlands in the Subarctic are largely caused by the presence of permafrost in soils, which significantly impedes soil drainage [Rodhe, 1990; Bartlett and Harriss, 1993]. Large volumes of organic matter are stored in these wetlands as peat deposits because low temperatures decrease rates of decomposition. The associated anaerobic environments are suitable sites for anaerobic bacterial activity. Methane emissions from these sources are controlled to a large extent by temperature, which causes seasonal CH_4 emissions [Williams and Crawford, 1984; Bartlett and Harriss, 1993].

By far the most extensive modern wetland areas are located in the tundra and boreal forest ecosystems of the high northern latitudes (Figure 7). These are largely found in Canada, Alaska, and northern Eurasia and consist of bogs that currently are major seasonal producers of CH_4. Most of these wetland systems developed during the Holocene (Table 4). The majority of modern bog areas at high latitudes [Matthews and Fung, 1987] were ice covered during the early parts of glacial terminations (Figure 7) when atmospheric CH_4 underwent rapid rises (Figures 1 & 2). Thus, these regions were clearly not available as a CH_4 source. Ice sheets effectively erode sub-glacial sediments, including any pre-existing accumulation of organic matter. Peat deposits would not have formed until well after initial deglaciation and ice sheet retreat in areas that were formerly glaciated. This includes areas of the continental shelf where ice sheets expanded to the shelf edge including the British Columbia and Labrador continental shelves.

Most of Canada was covered by ice sheets until major melting began at ~15 ka. It was not until ~7 ka that all of continental Canada became free of ice sheets [Zoltai and Pollett, 1983]. In the Laurentide region, ice sheet retreat was most rapid between 12 and 8 ka, and yet the interval of most rapid increase in the rate of peat accumulation did not occur until between 8 and 4 ka (Figure 13) [Harden et al., 1992]. Timing of initial peat formation varies throughout Canada depending on local climate and other conditions. In western and central Canada, deglaciation was completed by 9 ka. Initial peat accumulation, however, lagged behind deglaciation until between ~8 to 4 ka [Halsey et al., 1998; Zoltai and Vitt, 1990] (Figure 13). In western Canada (Alberta, Saskatchewan and Manitoba) peatlands began to form after ~9 ka, and then steadily increased during the Holocene, especially after 6 ka (Figure 14) [Vitt et al., 2000]. In northern Ontario and Manitoba, peat began forming after 5 ka, but in the Arctic region, peat formation was much more common from ~9 to 8 ka [Tarnocai, 1998], compared with the late Holocene, when little peat accumulation occurred. Cored sequences from the Canadian Beaufort Shelf show no evidence of any major pre-Holocene peat formation. Virtually all dated peat is of Holocene age [Hill et al., 1985, 1993]. In eastern Canada, wetland expansion occurred much later (~3 ka) [Liu, 1990].

Thus, it is clear that peat accumulation distinctly lagged deglaciation (Figures 13 & 14). The formation of peat is a complex process resulting from several interrelated factors including ecosystem development and climate [Korhola, 1995]. According to Vitt et al. [2000], this lag results from the time required to develop conditions conducive to peat accumulation, such as stabilization of seasonal water levels, restriction of water flow through the wetland, and sufficient leaching of soluble salts from the rock substrate to allow establishment and development of a moss layer [Zoltai and Vitt, 1990; Vitt et al., 2000].

A lag in peat accumulation following deglaciation is also evident in lower latitude areas to the south. In the area formerly occupied by Lake Agassiz, a vast pro-glacial lake in Manitoba and northern Minnesota, deglaciation occurred at ~11 ^{14}C kyr B.P. while peat formation did not commence until 4 ka [Wright and

TABLE 4. Timing of Wetland Development (Peatlands) in High Latitudes of the Northern Hemisphere. Major wetland expansion occurred during the Holocene especially in the middle to late Holocene.

Area	Age (ka)	Change	Reference
North America	~13	Minor peat expansion.	Gajewski et al., 2001
	<~9	Major peat expansion.	Gajewski et al., 2001
Continental Western Canada	<9	Peatland initiation.	Vitt et al., 2000
Western Central Canada	8 - 4	Peatland initiation.	Halsey et al., 1998; Zoltai & Vitt, 1990
North Central Canada	10 - 7	Peatland initiation.	Zoltai & Tarnocai, 1975
Canada, Laurentide Region	8 - 4	Peatland initiation.	Harden et al., 1992
Canada, Ontario/Manitoba	<5	Peatland initiation.	Zoltai & Pollett, 1983
Eastern Canada	<3	Wetland expansion.	Liu, 1990
Canadian Arctic	Mostly 9 - 8	Peatland initiation.	Tarnocai, 1998
Beaufort Shelf	<10	Peatland initiation.	Hill et al., 1985; 1993
U.S. Minnesota	<4.0	Peatland initiation.	Wright & Glaser, 1983
	<2.5	Peatland initiation.	Teller et al., 2000
			Romanowicz et al., 1995
Russia: Western Siberian Lowlands, Ob River	<12.4, Mostly Holocene	Peatland initiation.	Smith et al., 2000
Russia: Volga River	<5	Peatland initiation.	Kremenetski et al., 1999
Russia: Siberia	Mostly late Holocene	Peatland initiation.	Velichko et al., 1998
	<12	Oldest bogs.	Botch & Masing, 1983
	<8	Main peatland expansion.	Neustadt, 1984
Europe	~13	Minor peat expansion.	Gajewski et al., 2001
	~9	Major peat expansion.	Gajewski et al., 2001
Finland	10 - 7	Initial peat accumulation.	Korhola, 1995
	<8	Major peat accumulation.	Ruuhijarvi, 1983
British Isles and Ireland	Mostly Holocene	Peat accumulation.	Taylor, 1983
	>10	Rare and small-scale peat accumulation.	Taylor, 1983
Southern England: Dartmoor	<6	Peat accumulation.	Charman et al., 1999

Glaser, 1983; Teller et al., 2000] or ~2.5 ka [Romanowicz et al., 1995]. Before this time, the area was probably a wet meadow [Wright and Glaser, 1983]. In the British Isles, mire development was sporadic during the LGM and deglacial interval (16 to 10 ka) [Taylor, 1983]. Even in southern England (Dartmoor) in a modern wetland area south of the previous ice sheet margin, peatland development did not occur until after 6 ka [Charman et al., 1999]. Preservation and survival of peat in this region was limited because of an intense glacial climate.

A valuable overview of the latest Quaternary evolution of peatlands is provided by Gajewski et al. [2001]. Peatland development since the LGM was recon-

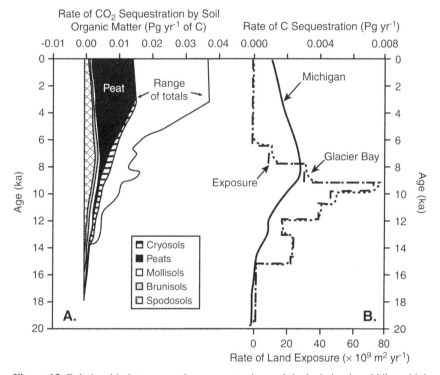

Figure 13. Relationship between carbon sequestration and deglaciation in middle to high latitude North American peatlands. (A) Model C sequestration rates for various soil types (see legend) in the region previously occupied by the Laurentide Ice Sheet in North America [Harden et al., 1992]. Almost all peat formed during middle and late Holocene. Peats have dominated soil C sequestration since the middle Holocene. In the Laurentide region, soil C sequestration increased most rapidly between 8 and 4 ka, following the most rapid ice retreat (12 to 8 ka). (B) Comparison of modeled rates of C sequestration and land exposure during deglaciation in Michigan and Glacier Bay, Alaska [Harden et al., 1992]. Differences in the rate of C sequestration in the two regions were tied to organic carbon decomposition dynamics. In Michigan, C accumulation lagged behind land exposure. In southern Alaska, C sequestration rates closely matched the rate of land exposure.

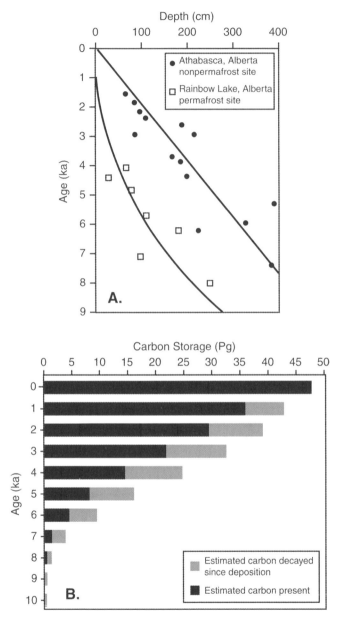

Figure 14. Timing of peat accumulation in western Canada [Vitt et al., 2000]. (A) Relationship between peat depth and calibrated, basal radiocarbon ages in the Holocene for two areas in Alberta. No basal peat ages are older than 8 ka and ~80% are younger than 6 ka. (B) Estimated changes in peatland carbon storage during the Holocene. Note that most C storage as peatland occurred after the middle Holocene (~5 ka).

structed over high and mid-latitudes of the Northern Hemisphere using the distribution and abundance of *Sphagnum* spores recorded in lake and bog sediment sequences. *Sphagnum* moss is by far the dominant component of bogs and poor fen peat, its remains making up most of the bog biomass. Since *Sphagnum* spores are wind blown, they represent a reliable proxy for presence of peatland over large regions. These reconstructions demonstrate that peatlands were essentially absent in North America and Eurasia during the LGM (Figures 15 & 16), with the exception of Alaska. Little carbon was stored in North American and Eurasian peatlands before the Holocene (Figure 17). *Sphagnum* peatlands began to develop in a very limited way during deglaciation in both North America (Figure 15) and Eurasia (Figures 16 & 17). Peatland development in North America and Eurasia became noticeable only after 9 ka, with major expansion following 5 ka (Figures 15 & 16). This study represents strong support that wetlands at high and mid-latitudes in the Northern Hemisphere developed almost completely during the Holocene with the greatest expansion during the late Holocene (< 5 ka). These data also suggest that little carbon was sequestered in Northern Hemisphere peatlands until after 9 ka (Figure 17).

Methane-producing wetlands are, however, highly sensitive to change, such that a lowering of the water table by 20 to 30 cm is capable of converting tundra from a CH_4 source to a sink [Reeburgh and Whalen, 1992]. The depth of the water table is controlled, as in other areas, by changes in sea level and precipitation/evaporation ratios. These variables would have played a significant role in controlling the extent of peat formation in these regions. Therefore, it is necessary to independently examine paleoecological records of wetland formation in areas not previously covered by ice sheets. The three principal potential areas are northern Eurasia, especially the Siberian plains, mid-latitude areas south of the Laurentide ice sheet in North America, and continental shelf areas that have since been inundated by sea-level rise.

Almost all peat in Siberia is Holocene in age. The oldest bogs in northern Russia date from ~12 ka [Botch and Masing, 1983]. As elsewhere, the Holocene can be regarded as a time of great peat formation, resembling older episodes of extreme peat development related to coal accumulation [Neustadt, 1984]. This is illustrated by the almost entirely Holocene development of Chistik Peatland in western Siberia (Figure 18) [Neustadt, 1984]. During the early Holocene (9 to 8 ka), only ~1.5% of the Siberian region was boggy. Major, near-continuous expansion followed, with rates of accumulation of ~0.2 to 0.8 mm yr[-1] [Neustadt, 1984]. Average peat thickness is usually not greater than 3 to 4 m, but is up to 13 m in places [Neustadt, 1984]. Within this region, the West Siberian Lowland is the world's largest high-latitude wetland with a 1.8×10^6 km^2 forest-palustrine region covering nearly two-thirds of western Siberia, of which at least half consists of peatlands [Smith et al., 2000]. The total carbon content within these wetlands is enormous (~215 Pg C) [Botch et al., 1995] representing one-tenth of the world's soil carbon and nearly half of all northern peatlands [Smith et al., 2000].

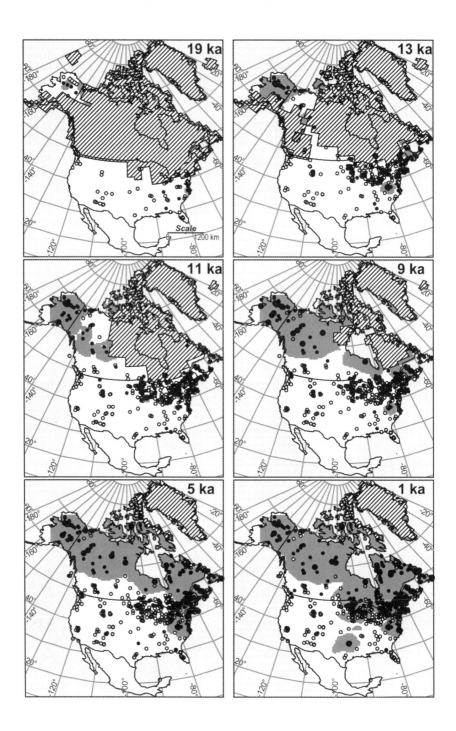

Studies in the Ob River region show that peatlands began to accumulate at 12.7 ka (based on basal radiocarbon dates), with most basal peat deposits dating to the Holocene [Smith et al., 2000] (Figure 19). No resemblance exists between the inferred history of this wetland development and that of atmospheric CH_4 during the late Quaternary.

Northern Eurasian wetlands have accumulated ~4 × 10⁵ Tg of organic carbon in sediments and peat [Zimov et al., 1997]. Melting of these frozen soils began sometime during the Holocene forming thaw lakes that now make up ~30% of the landscape and contribute CH_4 to the atmosphere [Zimov et al., 1997]. However, it appears that much of this system did not develop significantly until the late Holocene. Paleobotanical and paleocryological investigations [Velichko et al., 1998] have shown that the area of tundra and wetlands within the Boreal forest zone of northern Eurasia was much smaller during the early Holocene compared with today. Because of changes in wetland ecosystem dynamics, they suggested 86 to 44% lower CH_4 emissions from these wetlands and bogs during the early Holocene (and last interglacial (MIS 5e)), compared with the present. The increase in CH_4 emissions to near modern levels occurred during the late Holocene as the present systems developed. This interpretation is consistent with that of Blunier et al. [1995], who considered the CH_4 rise during the late Holocene to have largely resulted from wetland and peat development in the high northern latitudes, the most extensive modern wetland system.

In East European Russia, the Volga River region was a cold, dry steppe with birch thickets before 10 ka [Kremenetski et al., 1999]. This was followed by forest expansion in response to warmer climates. A major swamp region close to the Volga River did not form until the middle Holocene (~5 ka) [Kremenetski et al., 1999]. In Finland, the first peat began to form at 10.7 ka following ice sheet removal [Korhola, 1995]. More intensive accumulation began after ~8 ka [Ruuhijarvi, 1983; Korhola, 1995].

Age of Mid-Latitude Wetlands

Although limited in extent, wetlands occur in mid-latitudes, especially in the subtropics. Most of these are of coastal wetland type and, as such, developed largely in the middle Holocene in response to sea-level rise deceleration and stabilization. The development of most mid-latitude North American wetlands,

Figure 15. Maps of *Sphagnum* moss distribution in North America (based on distribution and relative abundance of spores) for six late Quaternary time intervals [Gajewski et al., 2001]. Open circles represent lack of *Sphagnum* spores for that time interval; small closed circles represent positive values less than 5% and large closed circles are values greater than 5%. Area with >0.5% is shaded. Locations with *Sphagnum* spores outside of the shaded area represent only traces. Ice sheet distribution is hatched. Note *Sphagnum* expansion largely occurred during the Holocene, especially in the late Holocene.

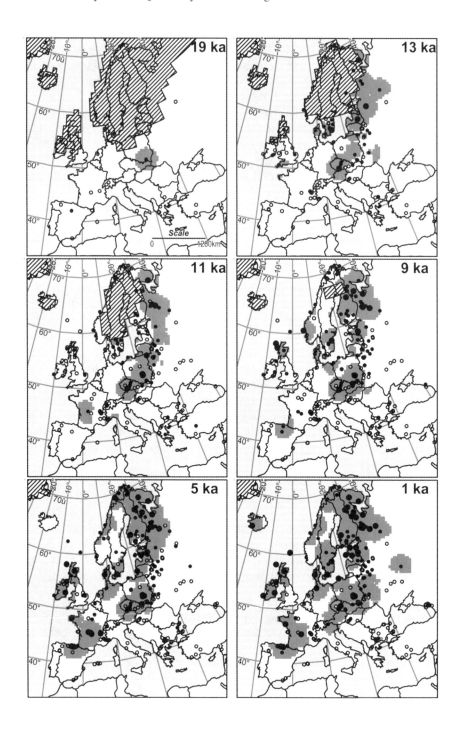

including those in the subtropics, occurred during the late Holocene (Table 5). Wetland formation in northern Florida began after ~6 ka [Watts and Hansen, 1988]. Investigations of peat accumulation in the Everglades, southern Florida, show that wetland development began at ~5.2 ka [Winkler et al., 2001] associated with a rise of the water table as a result of sea-level rise. In northwest Florida, a pluvial period of fluvial and karst infilling occurred between 9 and 4.5 ka [Chen et al., 1999]. Prior to the Holocene, the Mississippi River flowed within a deeply entrenched valley across the continental shelf and coastal plain [Fisk and McFarlan, 1955; Roberts and Coleman, 1996]. Highly organic backswamp clays did not begin to accumulate in this region until the late Holocene after significant valley filling had occurred in response to the base level change associated with sea-level rise [Fisk and McFarlan, 1955]. The marshes of the Louisiana-Mississippi wetland system began to form between 6 and 4 ka [Grimm et al., 1993], clearly as a result of adjustment to sea-level rise. The wetlands at mid-latitudes elsewhere in North America and Europe also mostly developed during the Holocene as reflected by the development of *Sphagnum* peatlands (Figures 15-17) [Gajewski et al., 2001]. The mid-latitudes appear to have been very dry before the Holocene except for a limited region in the northeast United States immediately south of the melting ice sheet during deglaciation (Figure 15).

Implications of Wetland Ages for Methane Sources

Considerable evidence thus suggests that neither high latitude nor tropical wetlands existed in any significant extent to support the large, rapid CH_4 emission rates at glacial and stadial terminations. This includes potential locations outside areas of modern wetland distribution. Atmospheric CH_4 increases were far too rapid (<300 yr) to have been produced by the expansion of wetland sources. The peatland systems were not sufficiently developed and widespread during the low sea-level stands, extensive ice coverage, and widespread aridity of the last glacial episode. Organic carbon buildup and peatland systems would not have been of sufficient extent for major CH_4 production. In the tropics, this system only began to form in association with prolonged seasonal flooding, and limited dry episodes. Major CH_4 production would therefore not be expected to respond immediately to rapid warming and inferred associated widespread increase in precipitation during the last glacial termination. Instead, the major modern CH_4-producing wetland systems would have required much more time to become established following the last glacial episode. Methane rise because of wetland

Figure 16. Maps of *Sphagnum* moss distribution in Europe (based on distribution and relative abundance of spores) for six late Quaternary intervals. Symbols are the same as in Figure 15. Note that *Sphagnum* expansion largely occurred during the Holocene [Gajewski et al., 2001].

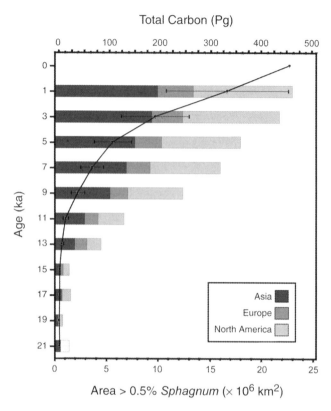

Figure 17. Changes in land areas supporting *Sphagnum* peatland (bars) since the Last Glacial Maximum for Asia, Europe, and North America [Gajewski et al., 2001]. Also shown is the change in carbon content of *Sphagnum* peatlands (line). Note major increase in peatland carbon accumulation during the Holocene.

activation should have been relatively slow, not rapid as indicated by ice core records. Furthermore, the late Quaternary sawtooth pattern of CH_4 variability is opposite to the pattern expected if wetlands were the primary source of atmospheric CH_4.

The large modern CH_4-producing systems did not develop until the late Holocene as a result of the hydrological changes associated with sea-level rise, removal of the Northern Hemisphere ice sheets, development of modern wetland ecosystems, and accumulation of organic carbon deposits, including peat. The rise in atmospheric CH_4 concentrations during the late Holocene (Figure 6) reflects the development of the modern wetland systems. The expansion of wetland systems during the Holocene associated with sea-level and water table rise, river channel in-filling, flood plain development, and ice sheet removal would have increased atmospheric water vapor, driving further atmospheric warming.

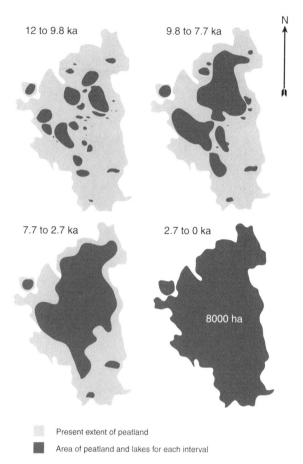

Figure 18. Development of the Chistik peatland, Western Siberia, through the Holocene illustrated in a series of four maps showing growth of the peatland and lake area to its present day extent of 8000 ha [after Neustadt, 1984]. Note major growth during middle to late Holocene.

This hypothesis differs from that of Ruddiman and Thomson [2001] who suggested that the major cause of the CH_4 rise over the last 5 kyr (Figure 6) prior to the industrial revolution was also anthropogenic. They suggested that 25% of atmospheric CH_4 input immediately prior to the beginning of the industrial revolution (~AD 1700) was anthropogenic in origin. While suggesting a dominantly anthropogenic origin for the middle to late Holocene rise in atmospheric CH_4 (primarily from rice cultivation), they reject an explanation based on expansion of boreal peat lands and tropical wetlands, and infer instead that major wetland systems had already developed by the middle Holocene. This inference contrasts with the large body of geologic evidence we have summarized supporting major

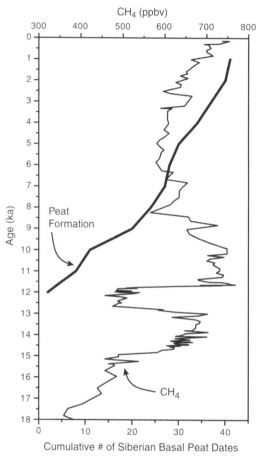

Figure 19. Comparison of latest Quaternary peatland initiation in the West Siberian Lowland (the world's largest high-latitude wetland system) [from Smith et al., 2000] with the GISP2 CH_4 record since 18 ka [Brook et al., 2000, and references therein] (see Figure 4). Peatland expansion is reflected by cumulative increase in number of basal peat ages (^{14}C) in 40 cores. Peat deposits less than 50 cm deep were not sampled, causing underestimation of recent peat initiation. Nearly all peatland development in this region occurred during the Holocene, well after major increases in atmospheric CH_4.

expansion of the wetland ecosystems at all latitudes during the middle to late Holocene. In spite of drier late Holocene conditions in many areas [Ruddiman and Thomson, 2001], major expansion of wetlands occurred in response to the development of favorable environments following deglaciation and mid-Holocene sea-level stability. It would be surprising if the dramatic expansion of major natural wetland systems during the Holocene did not lead to a significant increase in atmospheric CH_4. In comparison to the magnitude of this natural

TABLE 5. Timing of Wetland Development in Middle Latitudes of North America.
Wetland expansion occurred during the Holocene, especially in the middle to late Holocene.

Area	Age (ka)	Change	Reference
Louisiana-Mississippi Marshes	<6	Wetland initiation.	Grimm et al., 1993
Northwest Florida	9 - 4.5	Fluvial and karst infilling.	Chen et al., 1999
North Florida	6	Wetland initiation.	Watts & Hansen, 1988
Everglades, Florida	5.2	Wetland initiation.	Winkler et al., 2001
High middle latitudes	9 <~6	Minor peatland development. Major peatland expansion.	Gajewski et al., 2001

expansion, we suggest that anthropogenic effects on atmospheric CH_4 concentrations before the last several hundred years would have been relatively insignificant. Furthermore, major expansion of rice cultivation in Asia appears to have been limited to the latest Holocene, especially the last 1000 yr or less [Bray, 1986]. To bolster their arguments, Ruddiman and Thomson [2001] suggested that Holocene CH_4 trends over the last 5 ka are anomalous compared with other interglacial episodes during the last 420 ka as recorded in ice cores [Petit et al., 1999] (Figure 2). Specifically they suggest that the double CH_4 peaks of the early and late Holocene are unique to interglacials. The Vostok ice core record, however, indicates that of the several longer interglacial episodes during the last 420 ky, only one (MIS 5e; 125 ka), clearly exhibits a single CH_4 peak (Figure 2).

During the last glacial episode, ice sheets covered vast Northern Hemisphere regions currently vegetated (Figure 7), and increased global aridity led to the expansion of deserts and grasslands (Plate 2). There is general acceptance that the continental reservoir of organic carbon was significantly lower during the last glacial episode, contributing CO_2 to the ocean/atmosphere system [Sigman and Boyle, 2000; Adams et al., 1990]. Estimates of the magnitude of this change range between 450 to 1350 Pg C [Sigman and Boyle, 2000; Prentice and Sarnthein, 1993]. Prentice and Sarnthein [1993] estimated a change of 450 Pg C, which is close to values inferred from changes in benthic foraminiferal $\delta^{13}C$ ratio [Sigman and Boyle, 2000]. Much of the carbon accumulation following the last glacial episode was associated with the evolution of wetland systems in the Holocene, both at high and low latitudes. For example, tundra and forest soils in formerly glaciated regions alone account for a 400 Pg C increase in the continental reservoir [Schlesinger, 1990; Prentice and Sarnthein, 1993]. Present day peatlands at various latitudes account for 280-450 Pg C [Adams et al., 1990; Gorham, 1991].

It seems reasonable to assume that CH_4 production increased as the wetland carbon reservoir accumulated after the LGM, and that their respective development was intimately connected. This was almost certainly a slow process, since the continental wetland carbon reservoir did not immediately expand during deglaciation, but mostly accumulated during the Holocene. A major factor contributing to the magnitude of CH_4 production from continental wetlands is the amount of organic carbon (peat) accessible for methanogenesis [Cao et al., 1998]. The geological evidence indicates that sufficient organic carbon accumulations did not exist at the end of the last glacial episode to produce the abrupt increase in CH_4 at the glacial termination.

Implications of Inter-Polar Methane Gradients for Source

It has long been accepted that an increase in tropical monsoon strength during interglacials and interstadials created conditions necessary for a major low-latitude CH_4 source. Therefore, it is not surprising that initial investigations [e.g.,

Chappellaz et al., 1993] assumed that tropical wetlands were the dominant source for rapid atmospheric CH_4 increases during glacial and stadial terminations. However, a dominant tropical wetland source is not supported by studies of inter-polar CH_4 gradients reconstructed from polar ice cores [Dällenbach et al., 2000]. Modern pole-to-pole gradients of a few percent result from an uneven latitudinal distribution of sources and sinks because of incomplete atmospheric mixing between Northern and Southern Hemispheres. Average inter-polar differences in CH_4 are 14±4 ppbv during stadials and 37±10 ppbv during interstadials (Figure 5). Changes in CH_4 source strength are considered to be largely responsible for the CH_4 gradient. Only during the LGM was the gradient minimal (-3±4 ppbv) suggesting that northern and southern sources were about equal (Figure 5). In contrast, interstadials were marked by major relative increases from northern sources, an unexpected result, and not predicted based on paleoglacial history. For example, the large CH_4 increase at the B-Å came from both northern latitude and other sources, and CH_4 increases during interstadials resulted from increases in northern sources [Dällenbach et al., 2000].

These results suggest greater CH_4 contributions from northern sources rather than tropical latitudes. They also imply that the dominant CH_4 source during interstadials could not have been tropical wetlands, in spite of evidence of increased precipitation in tropical northern South America [Peterson et al., 2000] and increased tropical monsoon strength in southeast Asia [Wang et al., 1999b]. Clearly, the presence of major Northern Hemisphere ice sheets during the last glacial episode severely limited the potential of the high latitudes as a CH_4 source. The primary high latitude CH_4-producing wetlands of the present day were covered by ice during this interval. As discussed earlier, paleobotanical evidence suggests limited distribution of CH_4-producing wetlands in other middle to high latitudes not covered by ice; and unlikely to contribute sufficient CH_4 to cause the rapid rises observed in ice cores. These results therefore suggest that the CH_4 from northern sources must have had an origin other than continental wetlands.

5

Methane Hydrates as Atmospheric Methane Source

Since it appears unlikely that continental wetland systems were the source of the rapid atmospheric CH_4 increases at the glacial and stadial terminations during the late Quaternary, another source must be identified. The only other source that could rapidly supply large volumes of CH_4 is the marine sedimentary methane hydrate reservoir. We now discuss the possibilities of this source.

Methane Hydrates: Characteristics and Distribution

Methane hydrates (clathrates), the ice-like solid form of CH_4, are formed as CH_4 molecules captured within a cage of water molecules. Methane hydrates [Sloan, 1990, 1998; Kvenvolden, 1993; Henriet and Mienert, 1996, 1998; Kleinberg and Brewer, 2001; Paull and Dillon, 2001] form under conditions of low temperature, high pressure, and sufficient gas concentrations [Kvenvolden and McMenamin, 1980; Sloan, 1990, 1998; Dickens and Quinby-Hunt, 1997] (Figure 20A). Under conditions of different pressures, temperatures and gas concentrations and compositions, hydrates can form different recognized structures. For the dominant type of methane hydrate with all cage sites filled, the CH_4 concentration is 7.6 mol L^{-1} or a CH_4 to water volumetric ratio of ~164:1 when dissociated at standard temperature and pressure. This amount of CH_4 is an upper limit because hydrate cages are often not filled completely. Natural gas hydrates also form from other gases, but methane hydrates are the most abundant and widely distributed, and of greatest potential climatic importance. These substances, composed of a solid structure of gas and water, have variably been referred to as hydrates or clathrates in reference to their distinctive characteristics (composition and cage-like structure).

The generation of CH_4 by *in situ* microbial activity in marine sediments is widespread along the continental margins and in marginal marine basins, wher-

Methane Hydrates in Quaternary Climate Change
© 2003 by the American Geophysical Union
10.1029/054SP06

ever high sedimentation rates or other oceanographic conditions contribute to the preservation and burial of organic matter [Gornitz and Fung, 1994]. Methane hydrates form in continental margin sediments marked by relatively high organic carbon content (>1%) necessary for methanogenesis. The vast majority of continental margin sediments contain organic carbon concentrations sufficient for CH_4 genesis and the formation of methane hydrates [Gornitz and Fung, 1994]. In addition to microbial methanogenesis, thermogenic processes can be important for generating CH_4 at greater depths. Methane production from deeply buried organic matter can occur at temperatures of ~80° to 150°C [Claypool and Kvenvolden, 1983]. This gas may then migrate vertically through the sediments via faults contributing to the methane hydrate reservoir and/or associated free gas [Kvenvolden, 1995]. Given these major sources, it is not surprising that estimates of the amount of CH_4 carbon stored in the methane hydrate reservoir are large, even though estimated values vary widely, and remain uncertain and speculative [Kvenvolden 1988a, 1993; MacDonald, 1990b; Sloan 1990, 1998; Gornitz and Fung, 1994; Buffett, 2000; Kvenvolden and Lorenson, 2001].

Global estimates of the CH_4 content of methane hydrates range from as low as 0.5 Eg (1 Eg = 10^{18} g) to as high as 24 Eg [Kvenvolden and Lorenson, 2001]. An intermediate value of ~10 Eg is often used as a consensus value based on independently determined values by Kvenvolden [1988b] and MacDonald [1990a]. Such a volume represents ~3000 times the amount of CH_4 in the modern atmosphere [Buffett, 2000] and is the largest fossil fuel reservoir [Dickens et al., 1997]. This large reservoir of exchangeable carbon is stored as solid methane hydrate, with possibly equivalent or greater volumes of free CH_4 gas trapped below the hydrate zone [Dickens et al., 1997]. In all, and despite significantly different estimates of CH_4 carbon stored as methane hydrate and associated free gas, it is clear that this reservoir contains sufficiently large volumes of CH_4 to potentially play an important role in global climate change.

Figure 20. (A) Effect of water depth (pressure) and temperature on phase relationships between methane hydrate and dissolved or gaseous CH_4 [after Kvenvolden and Grantz, 1990]. Superimposed on this diagram (shaded area) is the range of southern California margin water temperature between modern (M) and Last Glacial Maximum (LGM) end members. A zone of potential methane hydrate instability during the late Quaternary (~400 to 1000 m) occurs in response to combined lower sea level (pressure effect) and temperature variation at intermediate-water depths. Methane hydrates are increasingly stable with increased water depth. (B) Schematic diagram illustrating potential distribution and thickness of methane hydrate zone in continental margin sediments [after Kvenvolden and McMenamin, 1980]. Stippled zone represents potential areas of methane hydrate formation, where pressure and temperature conditions combine to produce methane hydrate stability, assuming a sufficient supply of CH_4. Thickness of this zone increases with depth (pressure) and lower water temperature. Geothermal gradient is assumed to be 27.3°C/1000 m. Zone of potential instability during the late Quaternary is between ~400 and 1000 m water depth.

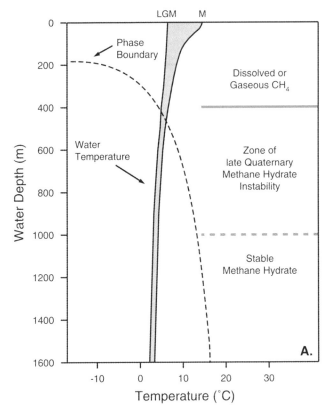

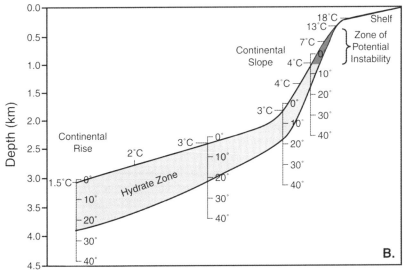

The formation of methane hydrates occurs within a stability field established by suitable pressure and temperature only when supply of CH_4 exceeds local solubility. Thus when sufficient supply of CH_4, is not available, hydrates do not fill the entire zone of potential stability [Xu and Ruppel, 1999]. In this case, the base of the methane hydrate does not necessarily coincide with the base of the methane hydrate stability zone. With sufficient CH_4 supply, a free gas zone may develop at some depth beneath the methane hydrate zone where temperatures are sufficiently high because of the geothermal heat gradient [Kvenvolden and McMenamin, 1980; Kvenvolden, 1993] (Figure 20B). If CH_4 supply is insufficient, a gap may develop between the base of the methane hydrate zone and the top of the free gas zone [Ruppel and Kinoshita, 2000].

Experiments have demonstrated that methane hydrate can form very rapidly in seawater [Brewer et al., 1997, 1999]. At low pressure, methane hydrate remains stable only near the freezing point of seawater, but at the greater pressures of increased water depth, hydrate can remain stable at temperatures as high as 20°C [Eyles, 1993] (Figure 20A). Pressure and temperature combine such that the water depth of potential methane hydrate instability, in areas outside modern polar regions, is between ~1000 and 400 m on the upper continental slope (Figure 20B). In polar regions, this zone is shallower because of very low bottom-water temperatures. The present location of methane hydrates in the ocean is directly determined by conditions of temperature and pressure (given sufficient supply of CH_4) that developed during the Holocene [Mienert and Bryn, 1997] following deglaciation and associated sea-level rise.

Sedimentary methane hydrate layers are mainly identified on the basis of their acoustic expression. The phase boundary between methane hydrate and free CH_4 can produce a prominent seismic discontinuity termed the bottom-simulating reflector [Holbrook et al., 1996]. The bottom-simulating reflector represents a boundary between a solid methane hydrate-rich layer above and gas-charged sediment below [Gornitz and Fung, 1994]. More accurately, the bottom-simulating reflector is the phase boundary marking the top of the free gas zone [MacKay et al., 1994]. Large volumes of free gas can be stored beneath the bottom-simulating reflector, especially in passive margin settings [Hyndman et al., 2001]. The extensive occurrence of methane hydrates has been inferred from the wide distribution of bottom-simulating reflectors [Kvenvolden, 1993; Kvenvolden et al., 1993]. Furthermore, it has now been established that methane hydrates can even occur over large areas in the absence of bottom-simulating reflectors [Holbrook et al., 1996; Holbrook, 2001; Dickens, 2001c] where supply is subcritical [Xu and Ruppel, 1999]. This suggests, in turn, that the volume of the methane hydrate reservoir may be much underestimated. Hydrates have now been sampled at ever-increasing numbers of locations including Blake Ridge, Hydrate Ridge, the margins off Peru, Costa Rica and British Columbia, northern Gulf of Mexico, Sea of Okhotsk and elsewhere [Kvenvolden and Lorenson, 2001; Spence et al., 2001]. Methane hydrates are widespread on both active and passive margins, on polar continental shelves, and in deep-water environments of inland lakes and seas [Kvenvolden and McMenamin, 1980; Sloan, 1990; Kvenvolden, 1993; Kvenvolden et al., 1993; Kukowski and Pecher, 2000]. Present on many continental margins [Kvenvolden and Lorenson, 2001], they are well known from the Arctic [Kvenvolden and Grantz, 1990; Majorowicz and Osadetz, 2001], circum-North Pacific [Kvenvolden, 1993; Hyndman et al., 2001; Ujiié, 2000], eastern equatorial Pacific [Kvenvolden and Kastner, 1990; Pecher et al., 2001], the Western Atlantic margin [Tucholke et al., 1977; Zonenshain et al., 1987], Gulf of Mexico [Brooks et al., 1986; Milkov and Sassen, 2000; Sassen et al., 2001a, b; Roberts, 2001], Caribbean [Kvenvolden, 1993] and other areas [Kvenvolden, 1993; Max, 2000; Kvenvolden and Lorenson, 2001; Kopp, 2002]. It is likely that the known distribution of methane hydrates on continental margins will expand with future surveys and syntheses. This is reflected by a recent review on the occurrence, geochemistry and formation of methane hydrates on convergent continental margins [Kastner, 2001]. In these regions, methane hydrates are commonly associated with clastic accretionary prisms of subduction zones [Hyndman et al., 2001]. Here, thick methane hydrate zones have accumulated by upward gas diffusion and advection in association with fluid expulsion in the accretionary prism [Hyndman et al., 2001; Ujiié, 2000]. In such environments, free gas volume appears unimportant relative to methane hydrates. Because hydrates accumulate as a result of upward fluid flow, the concentrations are not directly related to high organic carbon content which can be low (<0.5%) [Hyndman et al., 2001].

The methane hydrate reservoir, when present, typically occurs at varying depths within the upper few hundred meters of marine sediment on the continental margins [Kleinberg and Brewer, 2001]. The distribution of bottom-simulating reflectors suggests that most methane hydrates in the modern ocean are quite remote from the ocean floor, commonly occurring at depths in sediments of ~200 m below the sea floor [Bell, 1982; Paull et al., 1995]. Nevertheless, the hydrate stability zone is not an impermeable barrier to free gas migration and the CH_4 gas reservoir is thus not necessarily static [Gorman et al., 2002]. Gas can be injected upwards through the hydrate stability zone along pre-existing migration pathways before forming hydrate [Gorman et al., 2002]. These parts of the methane hydrate reservoir therefore become more vulnerable to instability due to climate or other changes. Where vertical CH_4 flux rates are sufficiently high through sediments, methane hydrates also can occur in sediments at or close to the ocean floor (<100 m sediment depth) and are thus more vulnerable to bottom-water temperature changes. These areas include Guaymas Basin, Gulf of California [Lonsdale, 1985], Santa Barbara Basin [Kennett and Sorlien, 1998], Gulf of Mexico [MacDonald et al., 1994; Sassen et al., 1999; Fisher et al., 2000], and Eel River Basin, northern California [Field and Kvenvolden, 1985; Brooks et al., 1991]. Areas where extensive deposits of methane hydrates are inferred to occur at shallow sediment depths include continental margins of the Arctic and

Subarctic, South China Sea, and Antarctic [Gornitz and Fung, 1994]. Methane hydrates were almost certainly more common and widely distributed during glacial episodes when bottom-water temperatures on the continental margins were significantly cooler.

The flux of CH_4 from marine sediments to bottom waters is modulated by anaerobic or sulfate-dependent CH_4 oxidation, a pervasive sedimentary process [Valentine and Reeburgh, 2000; Valentine et al., 2001, and references therein]. Such oxidation consumes most CH_4 in diffusion-dominated sediment [Reeburgh et al., 1993] and plays an important, although poorly quantified role in seeps and vents [Valentine et al., 2001]. Steep pore water sulfate gradients may reflect large upward fluxes of CH_4 from methane hydrates [Borowski et al., 1996] even when the existence of hydrates are not indicated by a bottom-simulating reflector. Methane discharge from ocean sediment also occurs as plumes extending upwards into the water column in various continental margin settings [Paull et al., 1995; Yun et al., 1999; Lewis et al., 1998; Suess et al., 1999]. Discharge of fluids is expedited by hydrate destabilization at depths that result in overpressurization by free gas that drives fluids upwards [Suess et al., 1999].

Methane that escapes sediment to the water column is further subjected to dissolution and then oxidation by aerobic methanotrophs (methanotrophy) [Suess et al., 1999; Valentine and Reeburgh, 2000; Valentine et al., 2001; Kvenvolden et al., 2001]. Biologically mediated CH_4 oxidation in the water column by methanotrophs can strongly control the amount of CH_4 that might reach the atmosphere from methane hydrate dissociation [Reeburgh et al., 1993; Valentine and Reeburgh, 2000; Valentine, in press]. Methanotrophy is crucial in limiting the residence time of CH_4 in the ocean. Residence time remains poorly constrained but has been estimated to be between 7 and 30 days [Kvenvolden et al., 2001].

The distribution of CH_4 within the water column is poorly known. However, undiscovered oceanic pools may be substantial [Sansone et al., 2001]. A significant subsurface pool of CH_4 was recently discovered in the eastern tropical North Pacific associated with the oxygen-minimum zone [Sansone et al., 2001]. The highest CH_4 concentrations within this pool (reaching 28 nM) are derived from upper continental slope sediments (300 to 600 m) off western Central America suggesting a methane hydrate source. Methane input into this pool from the sediments may modulate the strength of the oxygen-minimum zone and thus the extent of this oceanic CH_4 reservoir. This pool represents a large source of CH_4 at shallow depths which may be readily accessible to the atmosphere through upwelling.

Methane release from ocean to atmosphere can occur when bubbles survive upward transport through the water column without dissolving. Thus transfers can occur in shallow water and during energetic releases from the ocean floor. For CH_4 transfer to the atmosphere, bubbles need to migrate upwards at least to within the upper ocean mixed layer, after which turbulence can transport dissolved CH_4 to the sea surface and then to the atmosphere. CH_4 gas bubbles rising from the deep sea may be coated with methane hydrate [Brewer et al., in press] or oil [MacDonald et al., 2002]. This coating reduces diffusion of CH_4 out of the bubbles thus enhancing bubble longevity during ascent through the water column, increasing the flux of CH_4 to ocean surface waters and atmosphere [Kvenvolden et al., 2001]. Vertical transport of dissolved and undissolved (bubble) CH_4 through the water column is also enhanced by upwelling flow caused by the rising bubble streams themselves [Leifer et al., 2000; Leifer and Clark, in press]. Significant transfer of dissolved CH_4 to the ocean surface and atmosphere also occurs when deep waters ventilate through wind-driven oceanic upwelling [Valentine et al., 2001; Rehder et al., in press].

Although most CH_4 released into the oceans is oxidized within the water column by methanotrophic microbes, large expulsions during transient events potentially can transport high proportions of CH_4 to the ocean surface and atmosphere. Greatly enhanced upward flow of CH_4 occurs because much greater volumes of CH_4 produce dynamic bubble streams. This in turn causes strengthened upwelling flow, higher concentrations of dissolved CH_4, decreased bubble dissolution and increased CH_4 bubble size containing higher proportions of CH_4 [Leifer and Clarke, in press].

Methane hydrates are also widely distributed in non-marine environments in continental Arctic areas such as Siberia, the MacKenzie River Delta, and the north slope of Alaska [Sloan, 1990; Collett, 1993; Baulin and Danilova, 1984]. On the north slope of Alaska, methane hydrates are very extensive beneath most of the coastal plain province, with a thickness greater than 1000 m in Prudhoe Bay. The cold surface temperatures of the Arctic have led to associated accumulation of onshore surface permafrost and methane hydrates at depth. In Arctic permafrost regions, methane hydrates occur at subsurface depths from ~130 to 2000 m, although they commonly occur at ~500 m [Collett, 1993]. Large volumes of gas (1.4×10^{13} to 3.4×10^{16} m^3) are stored in these hydrates [Collett, 1993]. Continental methane hydrates may have contributed towards major increase in atmospheric CH_4 during glacial terminations as recorded in ice cores [Nisbet, 1992]. Kalin and Jirikowic [1996] have suggested that the rapid increase in CH_4 at the onset of the Bølling-Ållerød resulted from dissociation of methane hydrates in paleosols at northern high latitudes due to destabilization of glacial ice and initiation of groundwater advection.

Although evidence exists for some instability of the continental methane hydrate reservoir [Collett, 1993], it may have been relatively stable during much of the Quaternary. This is because of the significant depth of the methane hydrate zone and the general persistence of the capping permafrost over wide areas. Furthermore, response to surface warming is very slow, with surface temperature changes taking many thousands of years to affect deep subsurface temperatures in the ice-bearing sequences [Lachenbruch et al., 1982]. Nevertheless, the overlying permafrost contains CH_4 with concentrations of ~4.3 g m^{-3} [Rasmussen et al., 1993]. Melting of the more accessible permafrost can thus release CH_4 into the atmosphere [Hatzikiriakos and Englezos, 1994; Hostetler et al., 1999]. Indeed, effects of modern melting have been observed in satellite data as major, brief atmospheric plumes emanating from the Russian continental Arctic. These may reflect CH_4 released by breakup of permafrost and associated methane hydrates [Clarke et al., 1986].

Processes that Destabilize Methane Hydrates

Nisbet [1990] and MacDonald [1990b] have both argued that major destabilization of methane hydrates is central to the processes that end glacial episodes. Their hypothesis invokes dissociation of methane hydrates because of reduced hydrostatic pressure associated with sea-level fall and possibly warming of bottom waters [Summerhayes et al., 1979; Carpenter, 1981]. In their scenario, a sufficient volume of CH_4 was released into the atmosphere to cause global warming through various feedback processes. The dissociation of methane hydrates was implicated from evidence of massive sediment failure on continental slopes. Hydrostatic changes related to sea level, even on tidal time scales, are known to affect rates of emissions from marine gas fields [Boles et al., 2001].

An ~120-m sea-level lowstand during the LGM would have reduced hydrostatic pressure sufficiently to cause the gas hydrate stability zone to shallow by ~20 m, decreasing its thickness by several percent [Dillon and Paull, 1983]. Gas hydrates at shallow water depths would have been particularly vulnerable to dissociation. This suggested the hypothesis that glacioeustatic sea-level fall and increasingly large CH_4 releases into the atmosphere limit the extent of glaciation [Paull et al., 1991]. However, this hypothesis is not supported in detail by the CH_4 record in ice cores [Chappellaz et al., 1990; Lorius et al., 1990], which exhibits increasingly low, stable values of atmospheric methane into the LGM [Eyles, 1993]. On the other hand, large, rapid CH_4 increases at glacial terminations, which represent the largest warmings of the Quaternary, are consistent with the hypothesis of Paull et al. [1991] that major glaciation is limited by greenhouse warming resulting from major methane hydrate dissociation.

Methane hydrate stability at continental margin depths was probably more sensitive to temperature than sea-level changes during the late Quaternary [Buffett, 2000] (Figure 20A). A change in temperature of only 1°C is sufficient to offset the effect of a 100 m change in sea level, although the equivalent temperature change has been estimated to be as little as several tenths of a degree by Buffett [2000]. Nevertheless, it should be emphasized that increase in temperature alone does not necessarily lead to significant dissociation of methane hydrate, which like ice, requires sustained input of heat energy and thus prolonged temperature rise. Thermal effects on hydrate stability have not been seriously considered in previous late Quaternary investigations, probably because of the general assumption that bottom-water temperatures on upper continental margins were relatively stable. However, significant bottom-water temperature changes within the zone of potential hydrate stability (~400 to 1000 m) (Figure 20) have recently been discovered

[Kennett et al., 2000a; Hendy and Kennett, submitted], refuting this assumption. Thermal effects on methane hydrate stability are shown on a small scale by *in situ* studies of outcropping hydrates on the Gulf of Mexico slope (~540 m) [MacDonald et al., 1994; Roberts et al., 1999]. Here, seasonal gas venting from methane hydrate dissociation results from small changes (1.5°C) in bottom-water temperature. Likewise, in shallow waters (<300 m) of the Barents Sea, methane hydrate dissociation occurs in response to seasonal bottom-water warming [Long et al., 1998]. The process may have occurred on a larger scale during the widespread temperature changes associated with glacial-interglacial and stadial-interstadial oscillations. Shallow water areas where the hydrate stability zone is thin are potentially the most vulnerable to methane hydrate dissociation and CH_4 release to the atmosphere [Buffett, 2000]. The potential for major CH_4 release will be further enhanced when the base of the methane hydrate layer lies at shallow sediment depths, because significant destabilization and dissociation can begin at the base, rather than the top, of a methane hydrate layer. Formation of hydrates at much shallower depths and higher temperatures occurs on continental margins (i.e., off southern Australia) with the addition of high concentrations (up to 15%) of hydrogen sulphide to CH_4-dominated gas [Swart et al., 2001]. These hydrates would therefore be even more vulnerable to the marine temperature and pressure (sea-level) changes that occurred during the late Quaternary.

    Propagation of the thermal signal from bottom-water temperature change downward through the sediment to cause methane hydrate dissociation is potentially complex [Dickens et al., 1995; Ruppel, 1997; Dickens, 2001a, c]. The duration of the lag between bottom-water temperature change and onset of methane hydrate dissociation is related to the depth of the hydrate layer, sediment composition, and porosity, but the rate of propagation is not well known. The thermal signal of bottom-water temperature change may be transmitted through the sediment column fairly rapidly, on the order of ~20 m/100 yr [Revelle, 1983]. Bell [1982] employed a rate twice this value, suggesting that warming of Arctic bottom waters (~300 m water depth) could cause methane hydrate instability in the upper 40 m within 100 yr. Buffett [2000] considered that thermal diffusion through the upper 200 m of sediments may take ~1000 yr, which would represent a considerable lag within the context of late Quaternary climate change. This process appears to be poorly understood. For example, a low temperature anomaly (~0.5° to 2.9°C) is associated with a bottom-simulating reflector at depth in Atlantic margin sediments [Ruppel, 1997]. One of several explanations offered by Ruppel [1997] to explain this cool anomaly is a delayed response to bottom-water cooling during the last glacial episode (~15 ka), considered to be ~2°C lower than modern values [Labeyrie et al., 1987]. Heat transfer at the hydrate to free gas phase change can also shape thermal profiles and gradients in the sediment. It is likely that during the last glacial episode, methane hydrates formed at relatively shallow depths in the sediment and therefore were more accessible to bottom-water temperature changes, thus capable of responding relatively quickly. Harvey and Huang [1995] investigated heat transfer and methane hydrate destabilization processes in marine sediments by employing a coupled atmosphere-ocean model under different magnitudes of inferred future global warming. The results suggest that the modern distribution of hydrates could contribute limited CH_4 to amplify global warming. These experiments were largely concerned with modern methane hydrate distribution and conditions, and made simple assumptions about connectivity between hydrates, ocean and atmosphere. However, the experiments did not consider potential effects of free gas associated with methane hydrates [Haq, 1998a] or the potential of catastrophic processes that can expedite transfer of CH_4 from ocean sediments to the atmosphere.

The rate of the methane hydrate reservoir's response to bottom-water temperature increases was likely accelerated by mechanical sediment failure on upper continental slopes. Hydrate dissociation induced overpressurization at shallow sediment depths (<100 m) caused liquefaction, slope instability and the triggering of submarine landslides [Gornitz and Fung, 1994; Lerche and Bagirov, 1998]. The process of slope failure can be further amplified by the presence of methane hydrates in a metastable state below the usual base of the stability zone [Buffett and Zatsepina, 1999; Buffett, 2000]. This enhances the vigor of dissociation by increasing pore pressure disturbance [Buffett, 2000]. Slumping that might result from this process potentially brings gas-charged layers close enough to the ocean floor and/or opens up gas-escape conduits, thus expediting CH_4 transfer into the ocean/atmosphere system. Catastrophic release of CH_4 from the ocean floor is probably essential for transporting large amounts of CH_4 into the atmosphere, otherwise high proportions are dissolved within the water column and do not reach the surface [Nisbet, 1990]. Even on the shallow (65 m) continental shelves of Santa Barbara Basin, ~50% of the CH_4 emitted from sea-floor seeps dissolves during transit through the water column and thus does not directly reach the atmosphere [Clark et al., 2000; Leifer et al., 2000]. However, strong evidence exists for enhanced marine CH_4 emissions to the atmosphere in areas of coastal upwelling [Rehder et al., in press]. Coastal upwelling is likely important for transporting CH_4 to the ocean surface, shortening the time the gas resides in the water column subject to microbial oxidation, thus increasing CH_4 emissions to the atmosphere [Rehder et al., 2002]. This mechanism was surely important in enhancing transfer of CH_4 to the atmosphere during past major episodes of CH_4 hydrate dissociation in coastal upwelling regions along continental margins, the very regions of potential methane hydrate instability.

Evidence exists for major modern atmospheric CH_4 emission from methane hydrates. Large CH_4 plumes from islands in the Russian Arctic (~76°N) detected by satellites represent such eruptions resulting from dissociation of permafrost hydrates [Clarke et al., 1986; Nisbet, 1989]. These plumes were brief and large (100 km long; 30 km wide) and were observable because entrainment of hydrate, water, and clay caused condensation of atmospheric moisture.

Methane hydrate instability on the continental margins results from both changes in bottom-water temperature and sea level. The most vulnerable time for instability would have occurred when bottom water warmed during sea-level lowstands. Such change occurred on the continental margin of southern California [Kennett et al., 2000a; Hendy and Kennett, submitted] and probably over broad areas of the ocean floor.

Consistency of Atmospheric Methane Behavior with Methane Hydrate Source

A primarily methane hydrate source for the rapid atmospheric CH_4 increases at stadial and glacial terminations would have resulted in a different atmospheric response than if the CH_4 were derived from continental wetlands. The pattern of atmospheric CH_4 variation consistent with a methane hydrate source includes rapid increases exhibiting conspicuous peaks or overshoots followed by slower decreases. More CH_4 should be released to the ocean/atmosphere system during the early stages of dissociation marked by greater availability of unstable hydrate including that close to the sea floor. Once the readily available or most sensitive regions of the hydrate reservoir became depleted, further instability would have decreased leading to a decrease in atmospheric methane. Thus, the general saw-tooth pattern of atmospheric CH_4 associated with each interstadial is consistent with this hypothesis (Figure 4). Atmospheric CH_4 behavior during the early and middle Holocene [Blunier et al., 1995] is also consistent (Figure 6) as exhibited by the rapid rise at Termination 1B, peak values in the earliest Holocene, and decrease to mid-Holocene values, until the increase coincident with continental wetland development during the late Holocene (Figure 21).

The extreme rapidity of atmospheric CH_4 rise (decades to centuries) at stadial and glacial terminations is more consistent with a methane hydrate source than a continental wetland source. As discussed earlier, the development and activation of continental wetlands into a major atmospheric CH_4 source is both slow and delayed. In contrast, methane hydrates can be dissociated rapidly, especially when close to the ocean floor. Furthermore, the association of rapid atmospheric CH_4 increases with brief overshoots at some stadial (Figure 4) and glacial terminations (Figure 2) suggests a dynamic input from a stored source rather than a passive response to environmental change. Methane release from wetlands in response to climate change is unlikely to have produced such overshoot events.

Potential Inconsistencies of Methane Behavior with Methane Hydrate Source

It has been suggested that large atmospheric CH_4 emissions from the methane hydrate reservoir should be recorded as conspicuous, brief fluctuations in ice cores, with multiple emissions recorded as multiple fluctuations [Nisbet, 1992; Thorpe et al., 1996, 1998]. Several workers have examined ice core records for an imprint of this type [Chappellaz et al., 1997; Brook et al., 1999, 2000; Blunier et

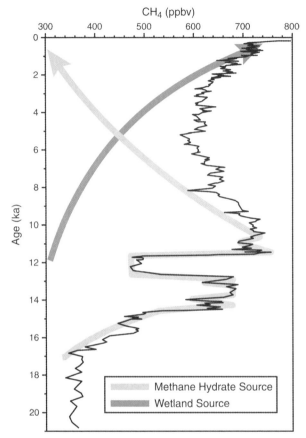

Figure 21. Schematic diagram illustrating suggested changeover in contribution to atmospheric CH$_4$ from methane hydrates and wetlands during and after deglaciation as recorded by GRIP CH$_4$ record from the Greenland ice core [Dällenbach et al., 2000] (see Figure 6). CH$_4$ from methane hydrates is proposed to have the characteristic sawtooth and spikey pattern of interstadials whereas wetland CH$_4$ emissions are inferred to reflect the history of Holocene development shown in Figures 13 to 19.

al., 1995] and concluded that this is unlikely to have occurred. The two main arguments against a methane hydrate source for the ice core CH$_4$ oscillations are: (1) the record is not sufficiently variable; and (2) CH$_4$ peaks are no higher than 750 ppbv [Brook et al., 1999]. Brook et al. [2000] argued that major (>50 ppbv) single events should be preserved without mixing, and yet are apparently absent. A modeling experiment of CH$_4$ increase at Termination 1 [Chappellaz et al., 1997] assumed only a methane hydrate source. In this experiment, the ice core CH$_4$ record could be reproduced by invoking a frequent or near-constant degassing of CH$_4$ from the hydrate reservoir, but not if all CH$_4$ was released in a small number of very large (4000 Tg) discrete events [Chappellaz et al., 1997]. Brook et al. [1999] and Blunier [2000] suggested that the ice core CH_4 records can only be explained by assuming a near-constant source of CH_4 during terminations, followed by maintenance of the source during the subsequent interstadials.

However, these models consider relatively infrequent, large episodes of CH_4 release. If methane hydrate accumulations with potential for dissociation were widespread in latitude, water depth, and depth in the sediment, which is the case, then the timing of smaller, individual expulsion events would be varied and incremental during a warming event. With a large enough number of abrupt, local events responding to different oceanographic and thermal propagation lags (with shallowest hydrate dissociation occurring first), the cumulative, global release of CH_4 could give the appearance of near-constant degassing. Only the largest individual expulsion events that distinctly exceeded the background flux to the atmosphere would stand out as distinct fluctuations in the ice core CH_4 record. Furthermore, the oceanic CH_4 reservoir [Sansone et al., 2001] may modulate transfer of CH_4 to the atmosphere from rapidly dissociating gas hydrates on upper continental slopes by acting as an intermediate reservoir for gradual release through oceanic overturning [Rehder et al., 2002]. This could have been a major process during episodes of methane hydrate instability such as during the glacial and stadial terminations and a contribution to the rapid development of the oxygen-minimum zone along continental margins [Cannariato and Kennett, 1999].

In spite of this criticism, the pattern of late Quaternary CH_4 variability does exhibit characteristics that would result from rapid methane hydrate dissociation, including: (1) the existence of rapid CH_4 increases at stadial and glacial terminations; (2) presence of CH_4 overshoots associated with some of these rapid increases, most likely only distinguishable where ultra-high-resolution analyses have been performed (see "Future Tests of the Hypothesis" chapter); and (3) intervals of short-term CH_4 variability. This is clearly noticeable during the early Holocene (>7 ka) as compared to the late Holocene (<7 ka), which exhibits distinctly smoother transitions in the GRIP record [Raynaud et al., 2000] (Figure 6). High-resolution marine climate records (Sargasso Sea [Sachs and Lehman, 1999] and Santa Barbara Basin [Hendy and Kennett, 1999; Hendy et al., 2002]) (Figure 3) exhibit large SST variability not reflected in the Greenland records. Such short-term climate behavior might be expected if forced by CH_4 emissions from methane hydrates. It remains unclear if the full magnitude of atmospheric CH_4 changes are recorded in the ice cores because of diffusional smoothing of the CH_4 while in the firn. Furthermore, signal smoothing may also be caused by atmospheric effects. Prather [1994, 1996] has suggested the possibility of hydroxyl radical (OH) feedbacks extending the atmospheric lifetime of CH_4 by several decades because of excess tropospheric CH_4 without requiring continuing fluxes into the system. Such episodes might occur during rapid emissions into the atmosphere, again smoothing the CH_4 record. This process may have been important during times of large CH_4 emissions to the atmosphere in the past [Schmidt and Shindell, 2001]. The potential effects of smaller emissions, such as occurred during the late Quaternary, remains to be determined.

Concern of an apparent absence of CH_4 increase associated with interstadials 19 and 20 [Brook et al., 2000], appears to have been at least partly alleviated. The Byrd ice core record [Blunier and Brook, 2001] shows small but distinct CH_4 rises in association with these interstadials. Furthermore, interstadial 20 clearly exhibits the classic sawtooth pattern for such events, with a rapid rise and gradual decrease in CH_4.

Inter-Polar Methane Gradients

The character of the inter-polar atmospheric CH_4 gradient reconstruction (Figure 5) is consistent with a primary methane hydrate source. Continental margins, including those with methane hydrates vulnerable to late Quaternary climate change are distributed in both hemispheres, but dominate the Northern Hemisphere. Furthermore, the behavior of Antarctic and Arctic continental margin methane hydrates should have been unequal. The methane hydrate reservoir along the Arctic margins likely exhibited more instability than the Antarctic reservoir, an area that experiences relatively stable water mass temperatures. This would also contribute to a skewing of CH_4 source toward the high northern latitudes.

Mass Balance Considerations

It is difficult to reconcile the deglacial and early Holocene CH_4 record with a chiefly wetland source of CH_4 when considering the history of wetland development. Mass balance models of the late Quaternary must consider which components of the CH_4 system were available at different times, not just which adjustments to the present system are required to reproduce ice-core measurements. Most modern non-anthropogenic CH_4 is derived from wetland sources [Ehhalt et al., 2001], but estimates of the relative contribution from tropical (~30 to 60%), temperate (0 to 15%), and high northern latitude (~30 to 60%) sources vary widely, with more recent estimates finding the greatest contribution (~60%) from the tropics [Bartlett and Harriss, 1993; Prather et al., 1995, 2001].

For example, Chappellaz et al. [1993, 1997] argued from mass balance calculations that the total amount of atmospheric CH_4 increase, and the measured inter-polar gradient, during the last glacial termination can be accounted for by approximately doubling the amount of wetland CH_4 production from the LGM to the Bølling. The total rise of CH_4 in GISP2 ice core from the LGM (350 ppbv) to the Bølling (675 ppbv) was 325 ppbv, reaching an atmospheric concentration approximately equal to the pre-industrial period (Figures 4-6) [Brook et al.,

1999]. Chappellaz et al.'s [1997] model tacitly assumes that wetlands at all latitudes during the LGM were sufficiently poised to very rapidly reach or exceed preindustrial levels of wetland CH_4 emissions in response to deglacial climate change. They concluded that wetlands of the middle to high northern latitudes contributed 25% of atmospheric CH_4 during the LGM and 33% of the doubled CH_4 levels of the earliest Holocene. This latter distribution of CH_4 sources is similar to that estimated for modern natural sources (60% tropical, 34% boreal) by Bartlett and Harriss [1993]; therefore they consider the earliest Holocene latitudinal distribution of CH_4 sources to be essentially the same as at present.

These model assumptions and results conflict with the geological evidence previously summarized that major high latitude wetland CH_4 sources did not exist during the LGM and had not significantly developed at the onset of the Bølling or even the earliest Holocene because of extensive ice cover, aridity, and insufficient time for peatland development (Figures 13-19; Plate 2). Without substantial contribution from a northern wetland component, tropical CH_4 production in the earliest Holocene would necessarily have been 50 to 100% greater than during the late preindustrial period. If the area of tropical wetlands during the Bølling or early Holocene was much smaller than at present, as was likely because of low sea level, better drainage and reduced floodplain extent, then tropical CH_4 production per unit surface area would have to be an unreasonable multiple of modern fluxes to account for the measured CH_4 values.

If, on the other hand, CH_4 emissions from sparse, patchy northern wetlands rose to 33% of global production during the early Holocene, reflecting a net increase from the LGM almost as great as that in the tropics (40 Tg yr[-1] versus 53 Tg yr[-1]) [Chappellaz et al., 1997], then the boreal wetland CH_4 flux per unit area would have been nearly an order of magnitude greater than today. The major high latitude wetland sources of CH_4, presently comprising 50% of the global wetland area, had clearly not yet developed at the onset of the Bølling because of extensive ice cover, yet CH_4 levels had nearly reached preindustrial levels. If this scenario is correct, the additional CH_4 contributions from these developing wetlands should have caused a much larger rise during the late Holocene than has been measured.

In any permutation, the *Wetland Methane Hypothesis* requires accepting geologically unsupported or unreasonable assumptions, namely: (1) CH_4-producing wetlands maintained similar latitudinal distributions from the LGM through the early Holocene, with global CH_4 flux chiefly modulated by proportional changes in the CH_4 production of each source region with only minor changes in their relative contributions; (2) If middle to high northern latitude wetlands were negligible sources of CH_4 in the earliest Holocene due to extensive ice cover and aridity, then tropical wetlands must have produced at least 50% more CH_4 than at present in spite of occupying much less area; or (3) middle to high latitude wetland CH_4 flux per unit area would have to be 5 to 10 times greater than today in order to contribute approximately the same amount of CH_4 to global production in the early Holocene, when geological evidence suggests that they occupied less than 20% of their present extent (Figures 14-19).

If northern wetland CH_4 sources were insufficient to account for the inter-polar CH_4 gradient and the global CH_4 rise, the potential contribution of emissions from methane hydrates on high latitude continental margins must be considered. If CH_4 from hydrates was dominant during the deglaciation and early Holocene, the mass balance considerations require this CH_4 source to have been completely replaced by those from wetland sources during the middle to late Holocene (Figure 21). Thus, methane hydrates would have stabilized while wetlands formed and expanded during the Holocene.

Holocene Change

Methane hydrates are thus implicated in the rapid atmospheric CH_4 increases at stadial and glacial terminations during the late Quaternary. This implies that the hydrate source that led to a doubling of atmospheric CH_4 at glacial terminations was later replaced by wetland CH_4 sources during the Holocene for reasons discussed above. Methane hydrates became increasingly stable during deglaciation and the early Holocene due to increasing pressures related to sea-level rise and stabilization of bottom-water temperatures after Termination 1B. As a result, hydrates have been relatively stable during the late Holocene and present day [Sassen et al., 2001a, b]. If this hypothesis is correct, it follows that the early Holocene atmospheric CH_4 maximum [Blunier et al., 1995] (Figure 6) was largely produced by methane hydrate emissions. Increasing stability and accumulation of hydrates from mid-Holocene to present, caused a decrease in CH_4 emissions before major modern terrestrial wetlands became fully developed. This scenario most likely contributed to the relatively low CH_4 concentrations during the middle Holocene. The CH_4 rise during the late Holocene (Figure 6) resulted from full-scale development of modern tropical and high-latitude wetland systems.

6

The Clathrate Gun Hypothesis

We now introduce the hypothesis that episodic atmospheric CH_4 emissions resulting from instability of the marine sedimentary methane hydrate (clathrate) reservoir contributed significantly to the distinctive behavior of late Quaternary climate on orbital (Milankovitch) and millennial time scales. Because this model involves punctuated releases of CH_4 from the methane hydrate reservoir, we have termed this the *Clathrate Gun Hypothesis*. According to this hypothesis, changes in upper intermediate waters intersecting upper continental slopes caused temperature changes of sufficient magnitude to partially destabilize the methane hydrate reservoir (Figure 22). Resulting CH_4 releases to the atmosphere/ocean system provided the amplification to "jump-start" rapid warmings at stadial and glacial terminations that were significantly reinforced by other greenhouse gases, especially water vapor. Collectively, these changes shifted the climate system into an interglacial or interstadial state (Figure 22).

According to this hypothesis, late Quaternary methane hydrate instability occurred even during times of relative sea-level stability because of frequent, rapid upper intermediate-water temperature oscillations over wide areas of the upper continental margins in the depth zone of potential hydrate instability. These temperature oscillations led to successive intervals of methane hydrate instability during the transitions and early portions of warm intervals and stability during cool intervals. It is suggested that the methane hydrate reservoir (clathrate gun) was episodically "loaded" or recharged during cold intervals of the late Quaternary when cold intermediate waters bathed upper continental slopes (Figure 22). The changing temperatures on the upper continental slopes resulted from oscillations in the production of upper intermediate waters in low and high latitudes (Figure 22). Switching to sources of warm intermediate waters at stadial and glacial terminations created instability in the methane hydrate reservoir and catastrophic release of CH_4 into the ocean/atmosphere system as a

Methane Hydrates in Quaternary Climate Change
© 2003 by the American Geophysical Union
10.1029/054SP07

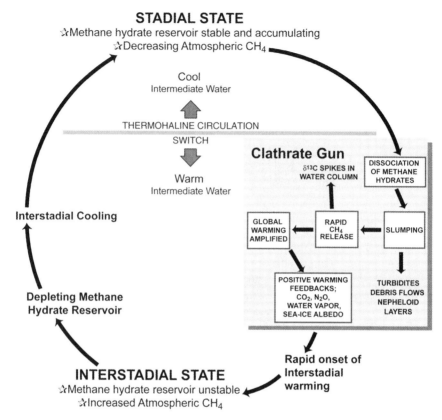

Figure 22. Summary diagram illustrating major elements of the *Clathrate Gun Hypothesis*. In this hypothesis, climatic feedbacks associated with changing atmospheric CH_4 composition result from changes in the stability of the methane hydrate reservoir associated with changes in bottom-water temperature (shifts in thermohaline circulation). Oscillations occur between stadial (glacial) and interstadial (interglacial) states.

result of sediment disruption that unroofed hydrates on the upper continental slopes. This led to the well-known rapid warmings of the late Quaternary on different time scales and magnitudes and also produced the sawtooth pattern of late Quaternary climate and atmospheric CH_4 variability exhibited in the 100-kyr cycle, Bond Cycles, and individual interstadials (Figures 1-5; Plate 1).

The hypothesis predicts extensive instability of upper continental slope sediments during rapid atmospheric CH_4 increases (Figure 22). This instability would have been reflected by widespread development of slumps, debris flows, and pockmarks on continental slopes, and associated mass sediment transport into the ocean basins as represented by turbidite and nepheloid layer deposits. The hypothesis predicts that glacial terminations represent intervals of greatest sensitivity of the methane hydrate reservoir because of vertical and lateral expansion of the zone of hydrate-bearing sediment due to prolonged cooling near the end of the glacial cycle combined with especially low sea level (and pressure). Methane emissions would have been greatest during these intervals when methane hydrates were destabilized, a prediction supported by the ice core records.

The oscillatory pattern of late Quaternary climate change at millennial to orbital time scales suggests a climate system highly sensitive to feedbacks within the system. We suggest that this sensitivity resulted from the changing interaction between Milankovitch-modulated and more rapidly changing intermediate-water distributions (both depth and extent) on the upper continental margins and sea level. This affected the relative stability of the methane hydrate reservoir and magnitude of CH_4 expulsions and associated climate change.

7

Methane Hydrates in Pre-Quaternary Climate Change

There seems to be widespread acceptance that changes in the methane hydrate reservoir played a role in climate history prior to the Quaternary. The origin of several brief episodes of global warming has been linked with massive dissociation of hydrates and CH_4 transfer into the atmosphere [Dickens et al., 1995, 1997; Dickens, 2000; Matsumoto, 1995; Retallack, 2001]. The strongest evidence implicating methane hydrates are records of large, negative carbon isotope excursions affecting carbon reservoirs of the ocean, biosphere, and atmosphere. Biogenic CH_4 has highly negative $\delta^{13}C$ values (~65‰) [Cicerone and Oremland, 1988] and therefore can provide a distinct isotopic fingerprint in sufficient environmental concentrations. Methane released from the hydrate reservoir is transferred to the exchangeable carbon reservoir via diffusion into the water column and/or outgassing into the atmosphere during sediment failure [Dickens et al., 1997]. Dickens [2001b] has suggested that the primary consequence of such CH_4 release may be dissolved O_2 depletion and carbonate dissolution in the ocean rather than atmospheric warming. The magnitude of the climate response would have largely depended on how much CH_4 reached the atmosphere. However, the intimate association between major climate warming and inferred CH_4 release points to a large increase in atmospheric CH_4.

Documented events based on large, abrupt negative $\delta^{13}C$ anomalies occurred during the latest Paleocene (55 Ma) [Dickens et al., 1995; Bains et al., 1999, 2000; Matsumoto, 1995; Kaiho et al., 1996; Katz et al., 1999], early Cretaceous (Aptian; 120 Ma) [Hesselbo et al., 2000; Jahren et al., 2001], late Jurassic [Padden et al., 2001], and early Jurassic (Early Toarcian; 183 Ma) [Hesselbo et al., 2000]. Major episodes of hydrate dissociation based on negative $\delta^{13}C$ anomalies in marine sequences have also been inferred for the Triassic/Jurassic boundary [Pálfy et al., 2001], the Permian/Triassic boundary [Krull et al., 2000; Krull

Methane Hydrates in Quaternary Climate Change
© 2003 by the American Geophysical Union
10.1029/054SP08

and Retallack, 2000; Twitchett et al., 2001; Musashi et al., 2001] (although other causes have been suggested to explain this anomaly [e.g., Ward et al., 2001]), the early Paleozoic [Quinby-Hunt and Wilde, 1995], and immediately following Neoproterozoic glacial episodes [Kennedy et al., 2001; 2002] (although see Hoffman et al. [2002] for alternate views). The late Paleocene episode, marked by rapid $\delta^{13}C$ shifts, began abruptly within 2 kyr, and lasted only ~220 kyr [Röhl et al., 2000]. It is inferred that CH_4, stored as methane hydrate, was released suddenly, then was gradually depleted from the ocean/atmosphere system, leaving a sawtooth shape to the $\delta^{13}C$ record. Dickens [2001c] compares the process to charging, discharging, and recharging an electrical capacitor. The early Jurassic event, estimated to represent one of the largest methane hydrate dissociation episodes of the last 200 myr, occurred mostly within only ~70 kyr [Hesselbo et al., 2000]. Both the late Paleocene and early Jurassic episodes were associated with oceanic warming, including deep waters that have been implicated in methane hydrate dissociation. We now describe the late Paleocene episode in greater detail because its detailed documentation provides significant insights about the potential role of methane hydrates in late Quaternary climate change.

The Late Paleocene Thermal Maximum was marked by a brief, intense episode of rapid global warming associated with a coeval massive perturbation of the global carbon cycle (Figure 23) [Kennett and Stott, 1991; Koch et al., 1992; Dickens et al., 1998, 2001b; Röhl et al., 2000; Zachos et al., 2001] and deep-sea faunal extinction [Kennett and Stott, 1991; Kaiho et al., 1996]. The perturbation is marked by a large, global negative excursion (~3‰) in the $\delta^{13}C$ of the ocean/atmosphere inorganic carbon reservoir (Figure 23). Dickens et al. [1995] and Matsumoto [1995] suggested that the negative $\delta^{13}C$ shift (Figure 23) reflects release and subsequent oxidation of large volumes (>1.1 to 2.1 × 10^{18} g) of carbon with a $\delta^{13}C$ signature of ~-60‰. The magnitude, shape and widespread nature of the isotopic anomaly (Figure 23) suggests sudden dissociation of methane hydrates at or near the onset of the event, followed by gradual return to the previous steady state condition [Dickens, 2001c]. About 2‰ of the excursion occurred within two steps which astronomical calibration suggests were each less than 1 kyr in duration [Norris and Röhl, 1999; Röhl et al., 2000], consistent with a series of massive releases of CH_4 from methane hydrates. The gradual return to earlier conditions occurred within 168 kyr [Röhl et al., 2000]. The changes in $\delta^{13}C$ and $\delta^{18}O$ thus describe a sawtooth pattern (Figure 23). The relatively rapid return of $\delta^{13}C$ to values similar to those before the CH_4 release may have resulted from an effective global feedback mechanism by which increased biological productivity and deposition of organic matter in the deep sea caused climate cooling because of the rapid removal of excess atmospheric CO_2 [Zachos and Dickens, 2000; Bains et al., 2000; Schmitz, 2000].

Methane hydrate destabilization during this episode is generally considered to have been caused by thermal diffusion resulting from rapid rise in deep-sea temperatures of 4° to 8°C [Kennett and Stott, 1991, 1995]. However, Katz et al. [2001] have suggested an alternative mechanism that involved unroofing of the

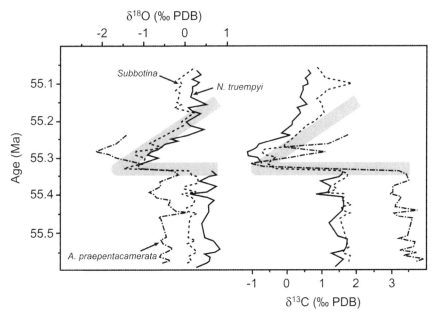

Figure 23. Dramatic negative oxygen and carbon isotope excursions marking the Late Paleocene Thermal Maximum, as recorded in benthic foraminifera (*Nuttalides truempyi*), inferred shallow-dwelling planktonic foraminifera (*A. praepentacamerata*) and inferred deeper-dwelling planktonic foraminifera (*Subbotina*) in Ocean Drilling Program Site 690, Antarctic Ocean [after Kennett and Stott, 1991]. Ocean warming (as indicated by $\delta^{18}O$) and inferred massive methane release from methane hydrates [Dickens et al., 1995] were most pronounced near the beginning of this brief episode, and then decreased, imparting a sawtooth pattern to the change.

methane hydrate reservoir through erosion by deep-sea currents on the western margins of the north Atlantic. In the thermal dissociation model, rise in pore pressure contributed to massive sediment failure and catastrophic release of large volumes of CH_4 into the oceans [Katz et al., 1999]. Evidence of extensive sediment failure on upper continental slopes is required to explain the Late Paleocene Thermal Maximum [Dickens, 2001c]. Indeed, a debris flow that may have formed during slope failure associated with inferred gas venting, occurred precisely at the onset of the carbon isotope excursion during the Late Paleocene Thermal Maximum as recorded in Ocean Drilling Program (ODP) Site 1051 on the Blake Plateau [Katz et al., 1999]. This is an intraformational clast layer made up of sediments transported from shallower depths on the continental slope of the Blake Plateau. Penecontemporaneous sediment failure of continental slope sediments took place along the slope of the New Jersey margin [Mountain, 1987], one of a number of episodes marked by slope failure and slumping. These caused slump-scar unconformities that Haq [1993] interpreted as down-slope movement of sediment blocks over lubricated horizons of destabilized methane hydrates. Unconformities attributed to this process have been reported from continental slope sequences on the New Jersey continental margin [Mountain and Tucholke, 1985; Mountain, 1987] and the Gulf of Mexico [Angstadt et al., 1983]. Haq [1993] suggested that mass sediment transport resulting from methane hydrate instability played a significant role in modifying sequence stratigraphic patterns on margins during sea-level lowstands.

Later we discuss evidence that suggests strong methane hydrate participation in late Quaternary climate change. If this was the case, we might expect similarities to exist in the behavior of the earth system during the Late Paleocene Thermal Maximum and the late Quaternary, even although models suggest different sensitivity of methane hydrates to climate change due to the fundamental differences in the oceanic state [Xu et al., 2001]. Although the magnitudes of changes are vastly different, there are features of the Late Paleocene Thermal Maximum that are matched in the late Quaternary. These are summarized briefly as follows:

- Evidence of large, rapid $\delta^{13}C$ excursions in sediment records suggesting a process of CH_4 outgassing through dissociation of methane hydrates.
- Inferred rapid rise in atmospheric CH_4 followed by a slower return to a previous steady state condition recorded directly by ice cores or indirectly by the isotopic composition of continental and marine carbonates.
- Close relation between inferred atmospheric CH_4 changes and climate.
- Episodes of rapid, major warming followed by gradual cooling imparting a sawtooth character to climate change.
- Close relation between deep-sea circulation modes and global climate change.
- Evidence of bottom-water temperature change at water depths of potential methane hydrate instability.
- Widespread evidence of massive sediment failure on continental slopes attributed to instability of the methane hydrate reservoir.

8

Primary Cause of Quaternary Instability of Methane Hydrates

Importance of Intermediate Waters

The depth of potential methane hydrate instability in the modern ocean is between about 1000 and 400 m on the upper continental slope (Figure 20B), except in the Arctic Ocean where the zone is ~200 m shallower. Phase changes in methane hydrates could be caused by changes in sea level (pressure) or bottom-water temperature or both (Figure 20A). In the world's oceans, the upper limit of methane hydrate presence, and hence instability, is usually limited by warm surface waters and low hydrostatic pressure. The lower limit of methane hydrate potential instability is set by depths where potential increases in water temperature or sea-level decreases do not intersect the phase boundary (Figure 20A). Marine geological evidence for slope instability, outgassing, and fluid flow in this depth range is discussed later.

Quaternary sea-level changes at stadial and glacial terminations were too slow to have directly caused rapid episodes of methane hydrate dissociation. Therefore, if the rapid climate warmings during the late Quaternary were triggered by CH_4 release from hydrates, there should be evidence for rapid temperature change of several degrees in upper intermediate waters. Furthermore, such temperature changes should have occurred in close association with stadial-interstadial and glacial-interglacial cycles.

Potential for Instability of Upper Intermediate Waters

Late Quaternary upper intermediate-water temperature changes have been neglected in determining relations between ocean circulation and climate change. Previously, little consideration was given to the idea that temperature

Methane Hydrates in Quaternary Climate Change
© 2003 by the American Geophysical Union
10.1029/054SP09

differences existed in intermediate waters on glacial-interglacial time scales. Instead, the major focus has been directed toward changes in middle to deep-sea thermohaline circulation [Broecker, 2000] within the context of climate change. However, evidence is emerging that upper intermediate waters also played a key role. Talley [1999] suggested that any model of present or past ocean circulation must include intermediate-water formation.

Evidence exists that significant intermediate-water temperature changes occurred on both glacial-interglacial and stadial-interstadial time scales. Intermediate-water temperature oscillations appear to have been closely linked to millennial-scale climate change on the California continental margin [Hendy and Kennett, submitted]. Changes in thermohaline circulation have strongly affected upper intermediate-water temperatures at the depths (<1000 m) of potential methane hydrate instability over large areas during the late Quaternary. Past changes in bottom-water temperatures have even been detected as temperature anomalies at depth in continental margin sediments [Ruppel, 1997]. These temperature changes almost certainly would have destabilized methane hydrates.

Modeling studies suggest expansion of intermediate waters within the Pacific during the LGM from both northern and southern high latitude areas [Ganopolski et al., 1998]. Significant expansions likely occurred in both North Pacific Intermediate Water (NPIW) from the north and Antarctic Intermediate Water (AAIW) from the south. Both of these source changes would have caused significant temperature changes at intermediate-water depths, even in the South Pacific, which is already influenced by AAIW at such depths. The timing of these expansions in the Northern and Southern Hemispheres may have differed according to the processes that changed the production rates [Ganopolski et al., 1998]. However coherent, decadal changes in salinity occurred synchronously in both NPIW and AAIW during the 20th Century suggesting that changes in these intermediate-water masses are linked interhemispherically [Wong et al., 1999].

We now summarize evidence of upper intermediate-water changes during the late Quaternary in association with their possible linkage to climate change. This evidence appears to be consistent with the *Clathrate Gun Hypothesis*.

Changes in Production of Intermediate Waters during the Quaternary

Modern processes that form North Pacific Intermediate Water. Paleoceanographic records suggest major oscillations in the production of NPIW occurred during the late Quaternary. We therefore review the processes involved in NPIW formation and how changes in these processes may have contributed to large production changes during the late Quaternary.

In the modern ocean, deep waters are not formed in the north Pacific because surface waters have low salinity and moderate density, even upon partial freezing (-1.8°C). North Pacific waters are fresh mainly because of the relatively low evaporation (55 cm yr[-1]) in this area compared with the North Atlantic (103 cm yr[-1]) [Weaver et al., 1999]. While limiting deep-water production, atmospheric and surface ocean conditions of the Northwest Pacific cause the production of NPIW, in an important modern process of potentially greater importance during cooler intervals of the last glacial episode. NPIW forms today in the Sea of Okhotsk [Freeland et al., 1998], and perhaps the Alaskan Gyre [Van Scoy et al., 1991] as a result of extreme cooling of surface waters. NPIW is detected in the upper 1000 m of the water column by a salinity minimum and relatively high oxygen concentrations compared with adjacent water masses [Weaver et al., 1999].

NPIW forms because of cooling and brine rejection during sea-ice formation and resulting sinking in the Sea of Okhotsk [Talley, 1991; Yasuda et al., 1996; Yasuda, 1997; Watanabe and Wakatsuchi, 1998], followed by tidal mixing and possibly deep convection in the southern Sea of Okhotsk [Talley, 1999]. A basin-wide 100 m thick layer of subzero (-1.5°C) water is formed centered at 150 m depth [Shiga and Koizumi, 2000]. The distinct signature of NPIW is formed through mixing in the region between the Kuroshio and Oyashio Currents [Weaver et al., 1999; Talley, 1999]. NPIW enters the North Pacific subtropical gyre through this strong interaction.

Although poorly known, the volume of modern intermediate waters flowing through the main passages from the Sea of Okhotsk into the Pacific has been esti-mated to be ~6 Sv [Takahashi, 1998]. Modern flow is relatively small because of low evaporation from intensely cool surface waters. Low SSTs are maintained because production of intermediate water is insufficient to draw in oceanic heat from lower latitudes and increase atmospheric temperature [Mix et al., 1999]. Thus, modern production of NPIW is low, largely limited to the Northwest Pacific gyre between ~300 and 700 m, and not found consistently across the Pacific basin in the broad California Current [Talley, 1993]. As a pervasive salin-ity minimum, NPIW is superficially similar to Labrador Sea and North Atlantic and Antarctic Intermediate Waters [Talley, 1993].

Oscillations in production of North Pacific Intermediate Water. Substantial evidence now exists indicating that major climatically-related oscillations in NPIW formation occurred during the late Quaternary. Isotopic data show increased production during glacial episodes [Curry et al., 1988, 1999; Duplessy et al., 1988; and Keigwin, 1998, and references therein]. Initial evidence from $\delta^{13}C$ changes suggested that NPIW expanded within high northern latitudes of the Pacific [Keigwin, 1998] and that its penetration was limited to the upper 2000 m of the Northwest Pacific. It is now known that well-ventilated NPIW expand-ed even further along the margins of the North Pacific during the LGM and other cold intervals, including the Younger Dryas [Keigwin and Jones, 1990; Behl and Kennett, 1996; Kennett and Ingram, 1995a, b; Ingram and Kennett, 1995; Emmer and Thunell, 2000; Hendy and Kennett, submitted].

The processes operating today could readily have been enhanced during cooler intervals increasing the production and extent of well-ventilated NPIW [Keigwin, 1998; Mix et al., 1999; Shiga and Koizumi, 2000]. This could have resulted from a change in evaporation minus precipitation at high latitudes or a reduction in Arctic freshwater discharge [Weaver et al., 1999], in turn related to extreme cooling of Northwest Pacific waters and increased sea-surface salinity driven by cold, dry winds from Asia [Mix et al., 1999]. Evidence exists for an oscillation in sea-ice extent in the Sea of Okhotsk between glacial and interglacial episodes that led to changes in NPIW production [Takahashi, 1998; Shiga and Koizumi, 2000]. During glacial episodes, sea-ice area increased on the western side of the sea, with the eastern side remaining open. This configuration increased the effectiveness of heat exchange with the atmosphere, enhancing ventilation and NPIW production [Takahashi, 1998; Shiga and Koizumi, 2000].

Oxygen isotope records indicate that significant temperature changes were associated with oscillations in upper intermediate waters on the California margin. At ~1000 m (ODP Site 1017), bottom waters warmed by at least 3°C close to the time of Termination 1A (Figure 24) [Kennett et al., in prep.]. Similar warming is recorded at ~500 to 600 m water depth in Santa Barbara Basin [Kennett and Ingram, 1995; Kennett and Venz, 1995]. Late Quaternary stadial-interstadial variability in NPIW production is inferred from benthic $\delta^{18}O$ changes recorded in Santa Barbara Basin [Hendy and Kennett, submitted]. Shifts of up to 0.75‰ occur between stadials and interstadials with low benthic $\delta^{18}O$ values during interstadials and high values during stadials (Plate 4). Intermediate-water temperature variability was as large as 3° to 5°C when inferred salinity differences are considered [Kennett et al., 2000a; Hendy and Kennett, submitted]. These oscillations appear to represent changes in the production of northern proximally-derived, cool, well-ventilated NPIW and southern distally-derived, warmer, poorly ventilated intermediate waters from low latitude Pacific sources (Figure 25) [Kennett and Ingram, 1995; Behl and Kennett, 1996; Keigwin and Jones, 1990; van Geen et al., 1996; Cannariato and Kennett, 1999; Zheng et al., 2000; Hendy and Kennett, submitted].

California margin records also suggest that surface and intermediate-water masses had different temporal responses or roles in rapid climate change. At ~1000 m on the upper continental slope (ODP Site 1017) off southern California, during the last deglaciation intermediate waters warmed ~700 yr prior to the major jump in surface water warming (Figure 24) [Kennett et al., in prep.]. Furthermore, in Santa Barbara Basin during stadial-interstadial transitions, intermediate-water warmings may have preceded the abrupt surface water warmings by 60 to 200 yr (Plate 4) [Hendy and Kennett, submitted]. This suggests earlier onset of warm undercurrent flow (300 to 500 m) associated with trade wind changes that preceded stadial and glacial terminations [Hendy and Kennett, submitted].

This millennial-scale variability of NPIW production led to major changes in ocean biological productivity [Berger et al., 1997], and oxygen-minimum zone strength along the California margin [Behl and Kennett, 1996; Keigwin and

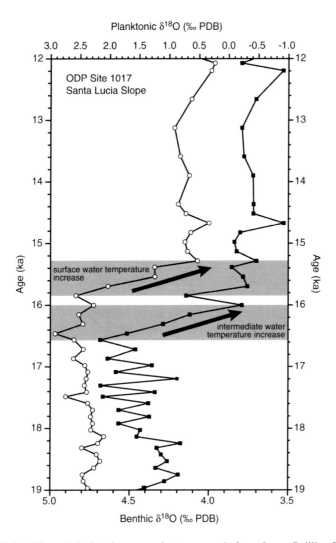

Figure 24. Benthic and planktonic oxygen isotope records from Ocean Drilling Program Site 1017 at 956 m water depth on the open-margin Southern California slope (Santa Lucia Slope) between 19 and 12 ka (last glacial through early deglacial episode) [Kennett et al., in prep.] (chronology from [Kennett et al., 2000c]). Temperature increases in both intermediate water (~3°C) and surface water (~8°C) are evident at the end of the last glacial episode. Most of the large decreases in δ18O in both benthic and planktonic foraminifera were too rapid to have resulted from ice volume effect on ocean isotopic composition. Furthermore, the negative δ18O shifts in both planktonic and benthic foraminifera do not co-vary, indicating their primary temperature origin. Note that the relatively rapid temperature increases in upper intermediate waters preceded that at the sea surface by several hundred years.

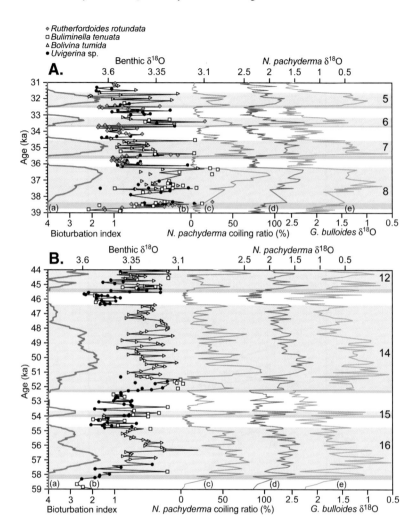

Plate 4. High-resolution time-series of millennial-scale climate change recorded in benthic and planktonic foraminifera and bioturbation history between 59 and 31 ka in Ocean Drilling Program Site 893, Santa Barbara Basin, California [Hendy and Kennett, submitted]. From left to right: bioturbation index (green) where 1 indicates laminated sediment facies and 4 indicates massive bioturbated facies [Behl and Kennett, 1996]; benthic foraminiferal $\delta^{18}O$ (black line average value); relative abundance of dextral to sinistral *N. pachyderma* (purple, shown as % dextral) [Hendy and Kennett, 2000]; $\delta^{18}O$ records of thermocline-dwelling *N. pachyderma* (blue), and surface-dwelling *G. bulloides* (red) [Hendy and Kennett, 2000]. Interstadials (light shading) are numbered according to the GISP2 scheme and defined by the planktonic response. Dark shading represents the time between benthic $\delta^{18}O$ decrease and indications of surface water warming (both faunal and geochemical evidence).

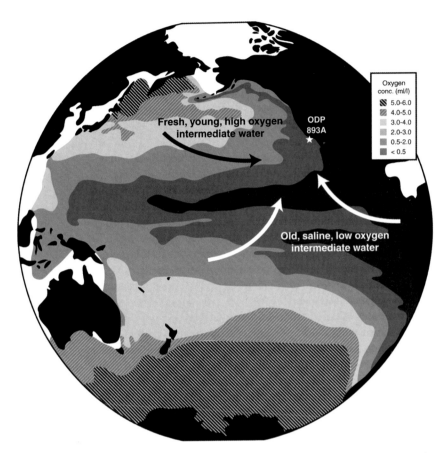

Figure 25. Modern oxygen concentrations (ml L⁻¹) of Pacific intermediate waters (27.807 στ (g L⁻¹) density surface [Reid, 1965]) reflecting competitive influences of northern (young, fresh, cold, high oxygen North Pacific Intermediate Water) and southern (old, saline, warm, low oxygen Southern Component Intermediate Water) sources.

Jones, 1990; van Geen et al., 1996; Cannariato et al., 1999; Cannariato and Kennett, 1999; Emmer and Thunell, 2000]. The oxygen-minimum zone was well developed during interglacial and interstadial episodes and poorly developed to absent during glacial and stadial episodes. This variability resulted from changes in deep and intermediate-water circulation as influenced by changes both outside [Broecker, 1991; Berger et al., 1997] and within the Pacific [Keigwin and Jones, 1990; Behl and Kennett, 1996; Hendy and Kennett, submitted]. These NPIW variations have special significance because of the resulting temperature changes within the depth zone of potential hydrate instability. The temperature shifts were of sufficient magnitude to cause methane hydrate instability and are thus consistent with the *Clathrate Gun Hypothesis*.

Antarctic Intermediate Water. The main thermocline of the modern ocean for much of the Southern and part of the Northern Hemispheres is occupied by inter-mediate waters that were formed in the Southern Ocean [Lynch-Stieglitz et al., 1994]. This AAIW forms near the Polar Front (Antarctic Convergence) in the circum-Antarctic region. A more dense component forms south of the Polar Front, with a less dense component (Subantarctic Mode water) forming in Subantarctic waters north of the Polar Front. AAIW forms a salinity minimum layer that descends to intermediate-water depths. A warm, salty component is added to AAIW from the northern Indian Ocean [Lynch-Stieglitz et al., 1994, and references therein].

Like NPIW, changes in production and processes of formation of AAIW are likely to have caused significant changes in temperature at intermediate-water depths (<1000 m) during the late Quaternary and hence are of potential impor-tance to methane hydrate instability. Little is known of this history, although it is clear that significant changes occurred in temperature and distribution of AAIW between the LGM and the Holocene [Lynch-Stieglitz et al., 1994; McCorkle et al., 1998]. These processes are likely to have affected upper continental margins in the Southern Hemisphere north of Antarctica and at low latitudes in the Northern Hemisphere. The formation of AAIW involves a temperature change that is about three times larger than that for Antarctic Bottom Water (AABW) formation [Talley, 1999]. If formation rates of AAIW and AABW were approxi-mately the same, AAIW would contribute more than AABW to global overturn-ing heat transport [Talley, 1999]. The modern northward reach of AAIW is large. AAIW is exported northwards into all three ocean basins (25°N in the Atlantic, 10°N in the Pacific, and 5° to 10°S in the Indian) [Talley, 1993]. AAIW proba-bly ventilates the equatorial thermocline in the Pacific [Toggweiler et al., 1991; Ninnemann and Charles, 1997].

Modeling studies suggest that AAIW production may have been much greater in the past [Ganopolski et al., 1998; Campin et al., 1999] expanding northwards across the equator to middle northern latitudes, ventilating intermediate waters (500 to 1500 m) [Mix et al., 1999]. Studies of Southern Ocean sediment sequences suggest potential changes in such sources [Ninnemann and Charles, 1997; Ninnemann et al., 1999]. Paleoceanographic investigations agree with modeling efforts indicating higher production and northward expansion of AAIW during the LGM [Lynch-Stieglitz et al., 1994; McCorkle et al., 1998]. In the North Atlantic, the northward reach was at least as far as northwest Africa at 20°N during the LGM. Here evidence exists for AAIW intensification between ~500 and 800 m leading to increased ventilation at oxygen-minimum zone depths [Sarnthein et al., 1982].

Lynch-Stieglitz et al. [1994] have conducted the most detailed study to date on AAIW change between the LGM and Holocene in southern Australian cores. They found that deep-sea temperatures below ~1200 m changed little between the LGM and Holocene. However, intermediate-depth cores exhibit significantly higher $\delta^{18}O$ values (0.2 to 0.6‰) than expected from the modern temperature profile. This indicates a relative cooling of ~1° to 3°C during the LGM. They further showed that glacial AAIW off south Australia is indistinguishable from Antarctic surface water. This suggests a much-reduced contribution to AAIW of warm, salty intermediate waters from the northern Indian Ocean. In effect, these changes indicate that AAIW during the last glacial episode was cooler, less saline, and better ventilated compared to modern AAIW.

These studies provide initial data suggesting significant changes in AAIW on glacial-interglacial time scales. It is likely that such behavior also occurred on millennial time scales in association with stadial-interstadial cycles. This remains to be recognized and documented.

Northern Indian Ocean Waters. Significant intermediate-water temperature changes shallower than 800 m have been recorded in late Quaternary sequences on the southern Pakistan continental margin. High-resolution studies reveal major changes in oxygen-minimum zone strength on orbital and sub-orbital time scales on both the Pakistan and Oman margins. These oscillations appear to correlate with millennial-scale climate oscillations as recorded in Greenland ice cores [Reichart et al., 1998; Schulz et al., 1998; Schubert et al., 1998; Altabet et al., 2002]. The oxygen-minimum zone was well established during interstadials leading to accumulation of organic carbon, and weak during stadials, with low organic carbon accumulation. The present day oxygen-minimum zone is maintained by both high productivity and by advection from the south of oxygen-depleted intermediate-water masses [Schulte et al., 1999]. Furthermore, intermediate waters were warmer during interglacials and interstadials and cooler during glacials and stadials [Reichart et al., 1998]. Two hypotheses have been formulated to explain these late Quaternary temperature oscillations. One suggests that major surface cooling, as much as 12°C during stadials, caused deep mixing to ~800 m leading to decreased productivity and disappearance of the oxygen-minimum zone [Schulz et al., 1998; Schubert, 1998]. This cooling was caused by strong, cold winter monsoon winds [Reichart et al., 1998] and thus was a top down process. The other hypothesis, instead, suggests a change in oceanic circulation [Schulte et al., 1999] with reduction in production of North Indian Ocean warm, saline waters [Lynch-Stieglitz et al., 1994] and associated northward expansion of AAIW during stadials with efficient ventilation at depths of the oxygen-minimum zone [Schulte et al., 1999]. The extent of temperature change at these intermediate-water depths (<800 m) remains undocumented but is likely to be large given the magnitude of SST change and because competing intermediate waters were derived from either the Antarctic or the subtropical northern Indian Ocean. In either case, large temperature oscillations at these depths have clear implications for methane hydrate instability on upper continental margins.

North Atlantic Waters. North Atlantic intermediate water also appears to have been involved in rapid climate change at both millennial and orbital time scales. Paleoceanographic evidence indicates intermediate waters changed over broad areas between ~350 and 1500 m water depth, with the clearest signals between 900 and 1000 m [Marchitto et al., 1998]. As in the North Pacific [Hendy and Kennett, submitted; Kennett et al., in prep.], intermediate waters (1286 m; Norwegian Sea) began to warm earlier in the deglacial period relative to the significant warming recorded in surface waters [Lassen et al., 2002]. During cold episodes, the thermocline became depleted in nutrients, while deep water became enriched [Slowey and Curry, 1995; Mulitza et al., 1998] and the oxygen-minimum zone disappeared during the LGM [Slowey and Curry, 1995]. These changes reflect rapid transformations in the relative strengths of NADW and overlying glacial North Atlantic Intermediate Water [Lynch-Stieglitz et al., 1999]. In the western North Atlantic subtropical gyre (Bahamas), glacial temperatures were 4°C cooler in the upper 900 m and 2°C cooler in deeper waters [Slowey and Curry, 1995]. These major temperature decreases in deep and intermediate waters resulted from cooling and significant sea-ice expansion in the high northern latitude source regions.

Changes in relative strength of North Atlantic intermediate and deep waters during the late Quaternary were probably related to changes in surface water salinity [Dokken and Jansen, 1999] (see Sarnthein et al. [in press] for review). Melt-water injections suppress convection at high latitudes in the North Atlantic, in turn causing rapid atmospheric cooling through reduction of NADW. In the Nordic Seas, intermediate waters are produced by brine release during stadials and through open-ocean convection during interstadials [Dokken and Jansen, 1999]. Major changes in circulation are recorded in the northeast Atlantic during glacial terminations. Clear evidence exists of warming of upper intermediate waters by different mechanisms, for example, beneath the Greenland Current during deglacial warming [Smith et al., 2001] and during the early Holocene under the influence of Mediterranean Outflow Water in the Gulf of Cadiz [Gardner et al., 2001]. Waters at ~1100 m depth in the northeast Atlantic became temporarily poorly ventilated during glacial terminations, probably because of convection shutdown associated with major surface melt-water influences [Venz et al., 1999]. Although the temperature effect of these ventilation changes on intermediate waters is unclear, they supply additional evidence of a dynamic history of intermediate waters in the North Atlantic.

Consistency with the Clathrate Gun Hypothesis

A central component of the *Clathrate Gun Hypothesis* is that intermediate-water temperature changes of sufficient magnitude occurred during the late Quaternary to episodically destabilize the methane hydrate reservoir on both orbital and millennial time scales (Figure 22). Although studies of intermediate waters (<1000 m) are limited, existing data indicate that significant intermediate-water temperature changes occurred during glacial-interglacial and stadial-interstadial transitions over large areas of the ocean and appear to be an important component of the rapidly changing climate system. Although sea-level changes may play an important role in the longer-term modulation of the methane hydrate reservoir and global climate [Haq, 1993, 1998a], these were not sufficiently rapid to cause the millennial-scale climate variability of the late Quaternary.

To summarize, studies of intermediate-water behavior during the late Quaternary indicate the following:

- Processes of intermediate-water formation at high latitudes were vulnerable to changes in state, which caused large production shifts during the late Quaternary.
- Modeling studies suggest significant increases in production of both NPIW and AAIW during the last glacial episode.
- Major late Quaternary environmental instability on millennial and orbital time scales occurred on the upper continental margin (<1000 m) as exhibited by dramatic oscillations in oxygen-minimum zone strength and significant intermediate-water temperature changes in the zone of potential methane hydrate instability.
- Large intermediate-water temperature changes over broad areas of the continental margins were associated with deglaciation.
- These appear to have involved sufficient temperature change to cause instability of methane hydrates.
- The high-resolution Santa Barbara Basin record exhibits bottom-water temperature oscillations associated with late Quaternary millennial- and orbital-scale climate variability.
- These appear to be associated with changes in production of upper intermediate waters at high and low latitudes.
- Intermediate-water warmings are closely associated and slightly lead the onset of rapid global warmings at stadial terminations.
- The magnitude of late Quaternary millennial-scale global climate change is closely matched by the upper intermediate-water temperature oscillations on the southern California margin during the last 60 kyr.
- Relative climatic and oceanographic stability marked both the Holocene and LGM compared to MIS 3. Instability during MIS 3 diminished towards the LGM.

The most profound global temperature rises occurred during glacial terminations, and were associated with large intermediate-water temperature increases. These intermediate-water temperature increases occurred during relatively low sea levels, potentially destabilizing the methane hydrate reservoir. Intermediate-

water temperature rise generally preceded (but did not lag) the onset of interstadials and the last glacial termination. This lead is consistent with the *Clathrate
Gun Hypothesis* because of the time required for methane hydrate dissociation
following bottom-water warming.

9

Instability of Methane Hydrates
During the Quaternary

We now review evidence that appears consistent with the following components of the *Clathrate Gun Hypothesis*: (1) significant instability of methane hydrates on the upper continental margins occurred during the late Quaternary; (2) methane hydrate instability contributed to major sediment failure and mass transport on continental slopes; (3) unroofed methane hydrates contributed to further hydrate instability and CH_4 release to the ocean/atmosphere system; (4) this process did not occur uniformly throughout the late Quaternary, but was focused at specific intervals.

Upward advection or diffusion of CH_4 through marine sediments is unlikely to be effective in transporting significant amounts to the atmosphere because the CH_4 flux from these processes is generally insufficient to overcome oxidation, consumption, or dissolution within the sediment or lower part of the water column. Catastrophic slope failure appears to be necessary to release a sufficiently large quantity of CH_4 rapidly enough to be transported to the atmosphere without significant oxidation or dissolution (Plate 5). The last major episode of widespread slope failure that likely contributed to this process occurred during the last glacial to interglacial transition. Information to support this comes from numerous continental margin sediment sequences.

Geochemical Signatures of CH_4 Release

Most CH_4 stored on continental margins as methane hydrate and free gas trapped beneath exhibits very negative carbon isotopic values ($\sim$-65‰) typical of biogenic CH_4 produced by methanogenesis within anoxic sediments [Kvenvolden, 1995; Cicerone and Oremland, 1988]. Release of sufficient quan-

Methane Hydrates in Quaternary Climate Change
© 2003 by the American Geophysical Union
10.1029/054SP10

tities of this CH_4 imparts a very negative $\delta^{13}C$ signal to the dissolved inorganic carbon of seawater through oxidation. This process has been described for the modern ocean by Suess et al. [1999].

Large, negative carbon isotope values have been recorded in benthic foraminifera of the last interglacial age from Peruvian margin near-surface sediments [Wefer et al., 1994] and cold CH_4 seeps on the northern California margin [Rathburn et al., 2000]. In both cases, the large, negative values appear to have resulted from environmental CH_4 influence, such as suggested by Borowski et al. [1999], rather than organic matter oxidation from sulfate reduction [cf. McCorkle et al., 1990] or enhanced organic carbon rain rate [Stott et al., 2002]. This interpretation was also made for large, millennial-scale negative $\delta^{13}C$ shifts in benthic foraminifera in the late Quaternary Santa Barbara Basin sequence (Plates 1 & 6) [Kennett et al., 2000a]. Millennial and orbital-scale oscillations in $\delta^{13}C$ varied by up to 5‰, being very negative (-2 to -6‰) during interstadials and the early Holocene and more positive (~-1‰) during stadials and the LGM. These oscillations were also inferred to reflect widespread shoaling of sedimentary CH_4 gradients and increased outgassing from methane hydrate dissociation during interstadials and the early Holocene [Kennett et al., 2000a] (Plate 5). We suggested that the magnitude of the benthic $\delta^{13}C$ oscillations are too large to be explained by vital effects or changes in oxidation or productivity. An alternate interpretation, that the $\delta^{13}C$ shifts could be due to oscillations in the organic carbon rain rate and consequent oxidation in the sediments, was proposed by Stott et al. [2002]. Intervals marked by large, negative $\delta^{13}C$ values, however, are marked by low diversity benthic foraminiferal assemblages of distinctive taxonomic composition, likely representative of anoxic conditions [Cannariato et al., 1999]. The taxa within these assemblages may be facultative anaerobes [Alve and Bernhard, 1995; Moodley et al., 1997, 1998], although the metabolic pathways through which the organisms survive anoxia remain obscure [Van der

---

Plate 5. Schematic diagram depicting three different ocean states and associated stability/instability of the upper continental slope methane hydrate reservoir. During glacial and stadial episodes (A), cold intermediate waters led to a relatively stable methane hydrate reservoir, in spite of lower sea levels. Greatest stability of hydrates occurred during the middle to late part of interglacial episodes (C) in response to high sea level and relatively long-term stability of intermediate waters. High instability of hydrates occurred during the intermediate deglacial times and following the onset of interstadials (B). Warming of intermediate waters then caused dissociation of hydrates. This in turn led to major instability of slope sediments, associated mass sediment transport and major release of CH_4 into the ocean/atmosphere system. The presence of methane hydrates on continental slopes is often detected seismically by the presence of a bottom-simulating reflector (BSR) if hydrates are underlain by sufficient free gas. The ocean states are marked by differences in vertical temperature gradient and changing strength of the oxygen-minimum zone (OMZ).

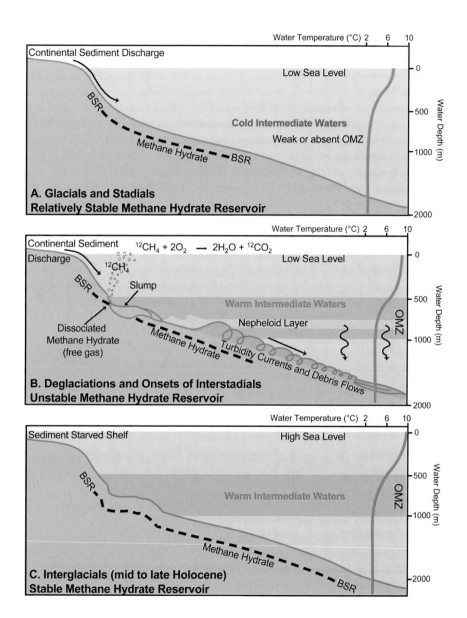

Water Temperature (°C) 2 6 10

Continental Sediment Discharge

Low Sea Level

BSR

Cold Intermediate Waters

Weak or absent OMZ

Methane Hydrate — BSR

Water Depth (m)
0
500
1000
2000

A. Glacials and Stadials
Relatively Stable Methane Hydrate Reservoir

Water Temperature (°C) 2 6 10

Continental Sediment $^{12}CH_4 + 2O_2 \longrightarrow 2H_2O + {}^{12}CO_2$

Discharge

$^{12}CH_4$

Low Sea Level

BSR

Slump

Warm Intermediate Waters

OMZ

Dissociated
Methane Hydrate
(free gas)

Nepheloid Layer

Methane Hydrate

Turbidity Currents and Debris Flows

Water Depth (m)
0
500
1000
2000

B. Deglaciations and Onsets of Interstadials
Unstable Methane Hydrate Reservoir

Water Temperature (°C) 2 6 10

Sediment Starved Shelf

High Sea Level

BSR

Warm Intermediate Waters

OMZ

Methane Hydrate

BSR

Water Depth (m)
0
500
1000
2000

C. Interglacials (mid to late Holocene)
Stable Methane Hydrate Reservoir

Zwaan et al., 1999]. The taxa that make up these assemblages are also known to be associated, although not exclusively, with modern CH_4 seeps [Wefer et al., 1994; Bernhard et al., 2001; Hill and Kennett, 2000, 2001] and hence Cannariato et al. [1999] interpreted their presence as additional evidence for episodes of high concentrations of CH_4 in Santa Barbara Basin bottom waters during the late Quaternary.

These inferred changes in benthic CH_4 flux in Santa Barbara Basin during the late Quaternary are supported by records of low $\delta^{13}C$ values of kerogens in basin sediments [Pratt et al., 1995]. These suggest episodic inputs of ^{13}C-depleted organic matter were produced either by methanotrophs [Hinrichs et al., 1999] or by chemosynthetic bacteria that utilize ^{13}C-depleted inorganic carbon derived from oxidation of CH_4. This record indicates periodic releases of CH_4 and episodic recycling of CH_4 in Santa Barbara Basin [Pratt et al., 1995] as has also been inferred from New Jersey shelf sediments based on the presence of highly negative calcite $\delta^{13}C$ values in sequences that are no longer experiencing methanogenesis [Malone et al., in press]. Even today, bottom waters in Santa Barbara Basin are more than 100 times saturated with CH_4 [Cynar and Yayanos, 1992]. Profiles of increasing CH_4 with depth in the reducing sediments of Santa Barbara Basin [Emery and Hoggan, 1958; Barnes and Goldberg, 1976; Warford et al., 1979; Doose and Kaplan, 1981] strongly suggest that basinal CH_4 is derived from the sediments. However, basinal CH_4 rapidly decreases to <10% saturation at ~100 m above the sea floor [Cynar and Yayanos, 1992] due to oxidation, consumption, and mixing.

Increased CH_4 flux likely resulted from methane hydrate dissociation triggered by rapid intermediate-water temperature changes at times of low sea level during the last glacial episode (Plates 1 & 4) [Kennett et al., 2000a; Hendy and Kennett, submitted]. The intermediate-water temperature changes were not random, but exhibit millennial-scale oscillations in close association with stadial-interstadial cycles, which are clearly recorded as changes in SST (Plate 4) [Hendy and Kennett, submitted].

Expulsion of hydrate-derived CH_4 into the water column from a convergent margin ridge off northwest North America [Suess et al., 1999] and in the Gulf of California [Merewether et al., 1985] generates submarine plumes hundreds of meters high and several kilometers wide. Suess et al. [1999] considered the release of such volumes of CH_4 to be of potential importance for climate change.

Buoyancy of released hydrates dislodged from the sea floor causes crystals to rapidly rise through the water column to the ocean surface [Brewer et al., 1999, 2000, submitted; Buffett, 2000]. Suess et al. [2001] reported the presence of large (1 m^3) fragments of floating and dissociating hydrate on the ocean surface off the Cascadia Margin (Hydrate Ridge) and suggest this as evidence of another mechanism for transport of CH_4 from the sea floor directly to the atmosphere. If of sufficient magnitude, this process has the potential to produce a negative $\delta^{13}C$ signal in the water column as the methane hydrate dissociates affecting surface waters remote from the ocean floor [Brewer et al., 1999, 2000, submitted]. This is otherwise difficult to achieve because most dissolved CH_4 emitted from the sea floor is oxidized at depth and does not reach surface waters.

Past episodes of major CH_4 release from hydrates into the water column were first inferred from $\delta^{13}C$ anomalies (Figure 23) [Kennett and Stott, 1991] recorded in the latest Paleocene for the entire ocean [Dickens et al., 1995, 1997]. Similar, but local, negative $\delta^{13}C$ anomalies have been identified in late Quaternary foraminifera [Maslin et al., 1997; Kennett et al., 2000a; Smith et al., 2001; Keigwin, 2002]. Brief, highly negative $\delta^{13}C$ anomalies recorded by latest Quaternary planktonic and benthic foraminifers in East Greenland (68°N) marine sediments have been attributed to CH_4 release into the water column from methane hydrate dissociation [Smith et al., 2001]. These episodes occurred during the glacial terminations at times of deglacial sea-floor depressurization resulting from ice sheet retreat and associated isostatic rebound, and bottom-water warming. Major CH_4 releases are indicated by $\delta^{13}C$ excursions in near-surface dwelling planktonic foraminifera living in the open ocean in close proximity to the East Greenland Current.

A late Quaternary sequence (818 m water depth) in the Guaymas Basin, Gulf of California, exhibits a distinctly negative $\delta^{13}C$ excursion in both planktonic and benthic foraminifera in the earliest Holocene (~9.5 ka) interpreted by Keigwin [2002] to represent a local major CH_4 release from dissociated methane hydrates affecting the entire water column. This $\delta^{13}C$ excursion is similar to those in Santa Barbara Basin [Kennett et al., 2000a] except that at this location, highly negative $\delta^{13}C$ values (-6‰) were recorded in an epibenthic foraminiferal species and not infaunal species. Keigwin [2002] interpreted these trends to reflect CH_4 release from a nearby location that was extreme enough to affect the $\delta^{13}C$ of the water column, but brief enough not to affect pore waters.

Distinctly negative $\delta^{13}C$ excursions (~-2‰) in planktonic foraminifera have also been identified in late Quaternary sequences of the Amazon Fan region [Maslin et al., 1997]. These are exhibited as brief excursions within the last deglaciation (15 to 13 ka) and have been interpreted to represent massive release of CH_4 from methane hydrates in Amazon Fan sediments. If so, these observations are especially significant because they are recorded by planktonic foraminifera that lived in open-ocean conditions marked by rapid surface currents. Methane injected into such active surface currents should be dispersed rapidly, and thus the negative $\delta^{13}C$ signals in planktonic foraminifera suggest major sustained sources of CH_4 from methane hydrates. The scale or duration of these events stand in contrast with the episodes recorded in the Santa Barbara Basin [Kennett et al., 2000a] and Gulf of California [Keigwin, 2002], much more restricted basinal settings.

During several interstadials, the Santa Barbara Basin benthic $\delta^{13}C$ record also experienced large, brief, negative excursions (up to 6‰) coinciding with smaller shifts (up to 3‰) in depth-stratified planktonic foraminiferal species (Plates 1

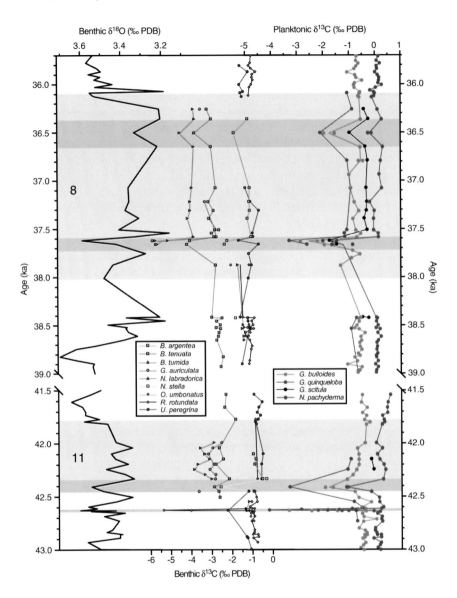

Plate 6. Detailed comparison of benthic δ¹⁸O and benthic and planktonic δ¹³C records in close association with D/O interstadials 11 and 8 from Ocean Drilling Program Site 893 (see Plate 1 for stratigraphic and paleoclimatic context) [after Kennett et al., 2000a]. Light shading represents laminated intervals. Dark shading highlights brief, negative δ¹³C excursions recorded in both benthic and planktonic species. Chronology same as that in Plate 1.

& 6). These intervals are inferred to record brief, massive CH_4 releases from methane hydrates changing the carbon isotope composition of the entire basinal water column [Kennett et al., 2000a]. Direct evidence for elevated bottom-water CH_4 concentrations during these events comes from the presence of the hopanoid diplopterol in the sediments, a biomarker particularly abundant in methanotrophic bacteria [Hinrichs, 2001; Hmelo and Hinrichs, 2002]. Correlation of the abundance of very negative $\delta^{13}C$ (-56 to -70‰) diplopterol with the $\delta^{13}C$ of benthic foraminifera in Santa Barbara Basin sediments indicates that the carbon isotopic signal recorded changes in CH_4 flux and bacterial oxidation of CH_4 in the basin's bottom water. Furthermore, the absence of lipids characteristic of microbes that oxidize CH_4 anaerobically within the sediment indicates that CH_4 was oxidized in the water column during these events (not in the sediments).

The massive CH_4 releases recorded in Santa Barbara Basin most likely resulted from deroofing of methane hydrates by slumping (Plate 5) [Kennett and Sorlien, 1998] and release of large volumes of free gas stored beneath the hydrate layer. Evidence exists in the basin for extensive mass wasting, mound fields, and pockmarks [Eichhubl et al., 2002]. Santa Barbara Basin slumps have head scarps containing slabs of authigenic carbonate likely representing evidence of past CH_4 expulsions [Greene et al., 2000]. Fluid continues to seep from slump scarps in the present day [Eichhubl et al., 2000]. The highly negative $\delta^{13}C$ values in surface-dwelling planktonic foraminifera could have resulted from accelerated upward transport of buoyant methane hydrates freed from ocean sediments [Brewer et al., 2000, submitted]. Several lines of evidence from biomarkers and the $\delta^{13}C$ of organic compounds indicate that even upper portions of the water column became temporarily anoxic due to depletion of oxygen by methanotrophy during individual brief episodes [Hmelo and Hinrichs, 2002].

Anomalous ^{14}C reservoir ages have been documented for marine waters at various times during the late Quaternary. Unexpectedly old ^{14}C dates were obtained from the last deglacial episode in several locations, including high latitudes of the North Atlantic [Waelbroeck et al., 2001], eastern Mediterranean [Siani et al., 2001], Cariaco Basin, Venezuela [Lin et al., 1997], and offshore of British Columbia, Canada [Kovanen and Easterbrook, 2002]). These results were interpreted as reflecting increased contribution of an older, deeper water component due to changes in the thermohaline circulation. Alternatively, these values could be produced by the release of ^{14}C-depleted CH_4 from methane hydrates into the water column, either locally or regionally, rather than by changes in ocean circulation. For example, X.-C. Wang et al. [2001] found from radiocarbon and carbon isotope investigations that hydrocarbon seeps contribute significant amounts of "old carbon" to the total organic carbon in sediments and deep water dissolved organic carbon pools in the northern Gulf of Mexico. They suggested that natural marine hydrocarbon seepage could be a significant source contributing "old carbon" to the marine environment with potential effects on radiocarbon ages in certain marine sediment sequences.

Timing of Instability of Continental Margin Sediments

The distribution of methane hydrates on upper continental slopes appears closely associated with evidence of mass sediment movement, pockmarks, and other features such as collapse structures [Dillon et al., 2001] resulting from mass sediment disturbance. Based on this relationship it should therefore be possible to indirectly infer timing of gas hydrate instability by determining the age of mass sediment disturbance. Mass sediment disturbance caused by methane hydrate instability is likely to be episodic, rather than continuous because of the different rates of change of sea level and bottom-water temperature related to climate oscillations of the late Quaternary (Plate 5).

The high-resolution carbon isotopic records from Santa Barbara Basin (Plate 1) [Kennett et al., 2000a] suggest modulation of methane hydrate instability by millennial-scale oscillations in intermediate-water temperature closely associated with stadial-interstadial cycles. Interstadials were marked by warmer intermediate waters that destabilized basinal hydrates and activated upward CH_4 flux through the sediment (Plate 1). Methane hydrate instability is also inferred to have caused sporadic submarine sliding in the basin that de-roofed methane hydrates and caused massive CH_4 release into the water column and atmosphere (Plate 5). In contrast, cooler stadial waters led to methane hydrate stability, accumulation, and reduced CH_4 flux into the basin (Plates 1 & 4). These oscillations were likely widespread along the California margin and elsewhere, affecting methane hydrate instability over broad areas of the margins, and contributing to millennial-scale atmospheric CH_4 oscillations [Kennett et al., 2000a].

This study implies that there has been a systematic history of major mass sediment deformation on upper continental slopes during the late Quaternary, i.e., slumping episodes activated by methane hydrate instability would have occurred at times of relatively low sea level and increased bottom-water temperatures (Plate 5). The last such major episode was during the last glacial to interglacial transition, an especially vulnerable time for methane hydrate instability. This should have been a time of major sediment instability on the slope. Although the age of most slump activation is poorly constrained, general trends of slump activity are now becoming known, and this record is summarized below (see "Future Tests of the Hypothesis" chapter).

Mass Sediment Transport

Submarine landslides are facilitated by a number of processes that cause sediment failure [McGregor and Bennett, 1979; Hampton et al., 1996]. One of the major processes now identified is related to methane hydrate instability. A strong association has been demonstrated between hydrate dissociation and instability of continental slope sediments (Table 6). Methane hydrates have a higher

TABLE 6. Late Quaternary Upper Slope Instability

Characteristic	Location	Reference
Continental margins where major Quaternary upper slope instability is associated with methane hydrates	General	McIver, 1977; 1982; Kvenvolden, 1993; Nisbet & Piper, 1999
	North Alaskan Slope	Kayan & Lee, 1991; 1993
	Nordic Seas	McIver, 1977; 1982; Hovland & Judd, 1988; Evans et al., 1996; Vogt & Jung, submitted
	Barents Sea	Mienert & Bryn, 1997; Bugge et al., 1988
	U.S. Atlantic Margin	Carpenter, 1981; Dillon, 1991; Dillon et al., 1991; Popenoe et al., 1993; Popenoe & Dillon, 1996; Paull et al., 1996; Rodriguez & Paull, 2000
	Mediterranean	Aloisi et al., 2000; Rothwell et al., 1998
	Gulf of Cadiz, Spain	Baraza et al., 1999; Gardner et al., 2001
	Southern California Margin	Kennett & Sorlien, 1998
	Pacific Margin	Kvenvolden, 1993
	Gulf of California	Lonsdale, 1985
	New Zealand	Lewis et al., 1998
	Southwest Africa	Summerhayes et al., 1979
Major slumps during deglacial and early Holocene	Wide-scale	Nisbet, 1992
	Norway, Storegga Slump	Bugge et al., 1988; Mienert & Posewang, 1999
	Barents Sea	Solheim & Elverhøi, 1993
	U.S. Atlantic Margin; Cape Fear Slide	Paull et al., 1996; Rodriguez & Paull, 2000
	Labrador Slope	D. Piper, personal communication
	Amazon Fan	Maslin et al., 1998
	Northern California; Humboldt Slide	Field & Baber, 1993; Gardner et al., 1999
Major slumps during Last Glacial Maximum	Eastern Mediterranean	Rothwell et al., 1998
Increased turbidite deposition during climate transitions	Madiera Abyssal Plain, NE Atlantic	Weaver et al., 1992; Rothwell et al., 1992; Masson, 1994
	Bengal Fan	Weber et al., 1997
	Gulf of Mexico, northwest slope	Badalini et al., 2000

mechanical strength than water ice (10 times stronger than ice at -13°C) [Haq, 1999]. This provides considerable strength to the hydrate-bearing sediment and can stabilize otherwise oversteepened slopes. During dissociation, however, pore fluid pressures become elevated within the methane hydrate layer relative to normal hydrostatic pressure. Even in the absence of hydrate dissociation, in areas of high sedimentation rates, excess pore pressures can develop in gas-charged porous sediment beneath the methane hydrate layer. Overpressurization in both cases will decrease the stability of slope sediments by reducing the normal effective stress acting on the sediment mass [Kayen and Lee, 1991]. Massive slope failure can result from decreased sediment strength, even on relatively low slopes [Coleman and Prior, 1988; Gornitz and Fung, 1994; Kennett and Sorlien, 1998].

Past methane hydrate instability is manifest on continental and insular slope sediments by slumps, slides, and other associated mass sediment movement, including turbidity currents, bench formation and collapse structures (Plate 5). Other disruptions of surficial sediments include the formation of pockmarks. A close relation between methane hydrates, bottom-simulating reflectors, and major submarine slumps has long been known (Table 6) [McIver, 1977, 1982; Hovland and Judd, 1988, and references therein]. This association seems to be widespread on oceanic continental margins [Haq, 1998a] including the North Sea [Hovland and Judd, 1988], Barents Sea [Mienert and Bryn, 1997], U.S. Atlantic margin [Carpenter, 1981; Popenoe et al., 1993; Dillon, 1991; Dillon et al., 1991; Paull et al., 1996; Popenoe and Dillon, 1996; Rodriguez and Paull, 2000], the Pacific margin [Kvenvolden, 1993], Gulf of California [Lonsdale, 1985], Mediterranean [Aloisi et al., 2000; Rothwell et al., 1998], the Alaskan slope [Kayen and Lee, 1991, 1993], southwest Africa [Summerhayes et al., 1979], New Zealand [Lewis et al., 1998], and other margins that have been surveyed in sufficient detail.

Other locations also show evidence of association between slope instability (slumps) and gas-charged sediments. In the Gulf of Cadiz, Spain [Baraza et al., 1999] slope instability features were identified on the upper slope between 140 to 270 m and 530 to 675 m. The slumps are 40 m thick and extend 1 to 8 km down the slope. Their association with gas charged sediment is indicated by zones of acoustic turbidity, gas-escape sea-floor craters and pockmarks, and acoustic anomalies in the water column indicating presence of gas plumes rising from the ocean floor. These features of sea-floor mass wasting and gas-escape have been correlated with increases in temperature associated with latest Quaternary changes in Mediterranean Outflow Water and dissociation of methane hydrates [Gardner et al., 2001].

The size and extent of many of the slides are impressive. For example, the Storegga Slide off Norway moved a total of 5600 km³ of sediment with a thickness of up to 450 m down the continental slope, apparently in three major episodes (one between 50 and 30 ka, and two in near succession between ~8 and 6 ka) [Bugge, 1983; Solheim and Elverhøi, 1993; Mienert and Posewang,

1999]. This and associated slides on the Norwegian margin may have been triggered by methane hydrate dissociation [Vogt and Jung, 2002]. Other major slides include the "Great Slide" south of Turkey [Woodside et al., 1998] and the Cape Fear Slide of the Atlantic U.S. province which is the size of Scotland [Evans et al., 1996; Popenoe and Dillon, 1996]. The Cape Fear Slide is one of many mass flow deposits that contribute to a total of 40% coverage of the surface of the North American Atlantic margin continental rise [O'Leary, 1996; Popenoe and Dillon, 1996]. Methane hydrates have been implicated for the origin of this slide. Methane hydrate thickness is reduced near the slide scar because of removal and dissociation [Popenoe and Dillon, 1996]. To the north off the New England margin, slumps originate at ~750 to 1000 m as indicated by location of head scarps [O'Leary, 1996]. Studies of the timing and spatial relations of mass sediment movement on the New England margin [O'Leary, 1996] indicate that most of the debris flows postdate canyon incision and associated sediment transport related to last glacial lowstand of sea level. Furthermore, the history of these debris flows appears to be independent of sea-level change and associated rates of sediment deposition on the outer continental shelf. These results instead suggest that the debris flows were formed by a process distinctly different and unrelated to those associated with changing sediment fluxes from the continents related to sea-level change. This process is more likely related to destabilization of the slope related to instability of the methane hydrate reservoir.

Landslides on the U.S. Atlantic continental slope exhibit a number of characteristics [Booth et al., 1993] consistent with an origin associated with methane hydrate instability:

- Steep slopes are not necessarily associated with more abundant slope failure. Most landslides occurred on slopes of 4° or less, at angles much less than normally required to influence slope stability under static conditions. Low angle landslides tend to be larger in area than those on steep slopes. These data are consistent with a methane hydrate origination process for landslide production unrelated to slope angle.
- Most landslides are disintegrative in character indicating that transient forces played significant roles in production of many slope failures. Booth et al. [1993] implicated both earthquakes and methane hydrates in this process.
- Origination depth of the greatest number of mapped landslides is between ~550 and 1300 m, with the mid-slope region between 800 and 1000 m being especially common. These would have been optimal depths for intermediate water-influenced methane hydrate dissociation during the late Quaternary. In a more restricted region of the New Jersey margin, McAdoo et al. [2000] found that slope failure occurred at greater water depths (1.4 to 2.6 km).

Processes causing slope instability were examined by McAdoo et al. [2000], by conducting morphometric analyses of submarine landslides in four distinctly different tectonic margin settings around the United States. They found that slope angle and proximity to seismic centers were unimportant factors in determining where failure occurred. Far more important were sediment rheology, sedimentation rates, and the presence of unconsolidated and possibly over-pressurized sediments. Most (75%) slides are disintegrative rather than cohesive suggesting failures involving significant amounts of energy and catastrophic beginnings. Large, extensive failures, typical of the Gulf of Mexico, are associated with high rates of sediment accumulation. The large size of many landslides on low slopes appears to have resulted from overpressurization and hydroplaning of debris flows, a process consistent with triggering by methane hydrate dissociation. McAdoo et al. [2000] found slope failures on the California margin evenly distributed between ~ 0.8 and 2.8 km water depth and those on the Gulf of Mexico slope between ~0.6 and 2.6 km.

Sequential slope destabilization can result from methane hydrate dissociation [McIver, 1982]. Methane hydrates at shallow depths in the sediment initially dissociate releasing CH_4 and increasing pore pressure. This causes bulging, venting, liquefaction, pockmark formation and slumping (Plate 5). Slumps then remove the sediment blanket and cause further dissociation driven by steeper CH_4 concentration gradients between hydrates and bottom water [McIver, 1982; Bratton, 1999]. Methane hydrate layers thus can become unroofed and fragmented (Plate 5). This can cause catastrophic release of free CH_4 previously stored beneath hydrate layers, contributing significantly to atmospheric release and potential climate effects [Dillon et al., 1991].

Sculpturing of the surface of Hydrate Ridge (~600 to 800 m water depth), offshore Oregon, was strongly influenced by the presence of methane hydrates. An unusually flat and shallow accretionary ridge with numerous scarps, benches, and slump deposits on its flanks reflect the strengthening of surficial sediments by methane hydrates and related precipitates [Tréhu et al., 1999]. The sea floor is covered by blocky carbonate and hydrate crusts. The stability of oversteepened ridge flanks was repeatedly undermined by fluid overpressurization beneath the impermeable hydrate cap, thus leading to slope collapse [Tréhu et al., 1999]. The dynamic evolution of the ridge involved repeated and interrelated episodes of formation and dissociation of methane hydrate and mass wasting. A history of methane hydrate instability is also recorded by the sequence of authigenic carbonate cements on the ridge [Bohrmann et al., 1998], exhibiting clear evidence of episodes of both stable and unstable states.

Timing of Slumps and Debris Flows

It has been long postulated that mass sediment movement down continental slopes was much greater during glacial episodes than during interglacial episodes, including the Holocene [Kennett, 1982]. Mass sediment transport down continental margins has been much reduced during the Holocene as illustrated, for example, by the Ebro margin in the northwestern Mediterranean Sea [Maldonado and Nelson, 1990; Nelson et al., 1991]. Higher rates of sediment transport during glacial episodes are represented by significantly higher rates of sediment transport down submarine canyons and continental slopes and by higher sedimentation rates in submarine fans as turbidites [e.g., Piper and Normark, 2001]. Much of this activity resulted from glacial low sea level that exposed continental shelves and estuaries to erosion and down cutting, and reduced the base level of continental erosion. Thus, glacial episodes are marked by increased terrigenous sediment input into the ocean basins compared with the Holocene [Francois et al., 1990; Piper and Normark, 2001]. Sediment supply also increased because of erosion by ice sheets and by extensive deforestation associated with glacial cooling and aridity.

However, superimposed on these glacial-interglacial differences is an apparent episodicity of mass sediment movement down continental slopes during the late Quaternary. This episodicity may have been enhanced by methane hydrate instability that assisted in destabilizing slope sediments. Widespread slumping and downslope mass sediment movement was not focused on the LGM when sea level was at its lowest [Paull et al., 1991], but rather during the glacial termination into the early Holocene when sea level was rising rapidly (Plate 5) (Table 6). Nisbet [1992] and Mienert and Posewang [1999] recognized that widespread scars of marine slumps were commonly formed during the glacial termination and in some cases are very large (Table 6). Activity of the Storegga Slump, off Norway, has been episodic during the late Quaternary. The first phase was between 50 and 30 ka within MIS 3 and the second and third phases during the early Holocene [Bugge et al., 1988]. On the eastern Reykyanes Ridge, at high north Atlantic latitudes, extraordinarily high depositional rates of siliciclastic sediments occurred during the last two deglacial episodes, the highest of any time in the late Quaternary [Ruddiman and Bowles, 1976]. Likewise, slumps off southwest Norway exhibit significant activity from the glacial termination into the early Holocene, in part activated by post-glacial regional isostatic rebound [Bøe et al., 2000]. The Barents Sea margin also exhibits evidence of methane hydrate dissociation during deglaciation after 15 ka [Solheim and Elverhøi, 1993].

Clear evidence for highest siliciclastic sedimentation during the deglacial through early Holocene interval (11 to 8 ka) was also found in sediment sequences on slopes and basins adjacent to the northeast Australian margin from Queensland to southwest Papua-New Guinea [Dickens et al., 2001]. At this time, a 1 to 2 m thick siliciclastic unit was deposited on the slope, reflecting an order of magnitude increase in siliciclastic accumulation. The timing of this major increase in sedimentation is unexpected because it coincided with rapid deglacial sea-level rise and associated increased accommodation space on continental shelves when sediment transport to the slopes should have decreased [Dickens et al., 2001].

The timing of episodic massive slumping is similar on the Amazon Fan. In this region, major latest Quaternary episodes have also been documented during MIS 3 (45 to 42 ka and 35 ka) and during the latest glacial through early deglaciation (17 to 14 ka) [Maslin et al., 1998]. The youngest deposit is capped by Holocene sediment, supporting a deglacial age of the event. These mass transport deposits originated on the continental slope between 200 and 600 m. The most recent activity of the Cape Fear Slide (~5000 km^2) off the North Carolina margin, an area with widespread methane hydrates, was during the glacial termination. Studies of piston cores suggest the most recent activity was between 14.5 and 9 ka [Paull et al., 1996]. At ODP Site 991, the slump unconformity is dated at = 10 ka [Rodriguez and Paull, 2000]. Off northern California, the Humboldt Slide [Field and Barber, 1993; Gardner et al., 1999] was most recently active during the late Pleistocene to early Holocene (before 10 ka). Considered by Gardner et al. [1999] to be representative of many slides on the continental slope, the Humboldt Slide is associated with a bottom-simulating reflector, abundant significant gas, and abundant pervasive pockmarks, suggesting that much of the surface vented during mass movement. The slide originated at 250 to 600 m water depth, within the zone of potential methane hydrate instability. These associations are consistent with a methane hydrate trigger for the slide, although they do not prove a causal relationship. Furthermore, on the Labrador Slope, greater mass sediment movement occurred during the glacial termination (15 to 12 ka) compared with the preceding glacial and following interglacial episodes [D. Piper, personal communication]. Two hypotheses have been proposed by Piper [personal communication] to explain this pattern: (1) The effect of regional warming on ice sheet melting, or (2) high sedimentation rates and compaction dewatering that destabilized upper slope sediments.

These data, although limited, suggest that increased slumping on the upper continental slope was not triggered merely by increased sediment delivery associated with the glacial low-stand maximum, but with processes that occurred over the glacial-interglacial transition, including the earliest Holocene (Plate 5). Inferred high rates of sediment transport may have resulted, in part, from remobilization and avulsion of sediments from the outer continental shelf during most rapid sea-level rise and shelf transgression [Dickens et al., 2001]. However, observations are also consistent with the hypothesis that relates increased mass sediment transport with methane hydrate instability through thermal dissociation (warming at glacial termination), rather than through decreased hydrostatic pressure (glacial low-stand maximum).

Timing of Turbidite Deposition

There is general acceptance that turbidite deposition in submarine fans, abyssal plains, and other basins peaked during sea-level lowstands and continued only through early sea-level rise, after which sediment-supply was arrested on the shelf as transgressive near-shore deposits [Vail et al., 1991; Weber et al., 1997]. Terrigenous sediment supply to the ocean basins was clearly greater during low sea levels. This was caused by increased river gradients because of entrenchment and canyon formation on shelves and slopes, and by shift of deltas to the shelf edge and increased slope instability [Prins and Postma, 2000]. Where delta fronts were positioned close to the shelf edge, turbidite deposition often continued during sea-level rise and highstand [Weber et al., 1997; Prins and Postma, 2000, and references therein].

Nevertheless, evidence is emerging that even higher rates of turbidite deposition occurred during the last glacial termination and other times of rapid climate change. High-resolution dating of the late Quaternary turbidite sequence from the Madiera Abyssal Plain, Northeast Atlantic, indicates that most turbidite layers were deposited during climate change at the boundaries of marine isotope stages, rather than during glacial episodes. Emplacement occurred during nearly all deglacial episodes since MIS 20 (706 ka) [Weaver et al., 1992; Rothwell et al., 1992; Masson, 1994]. Furthermore a dramatic pulse of sediment deposition during the glacial to interglacial transition has been documented on the Bengal Fan [Weber et al., 1997]. This began contemporaneously with Termination 1A and ended after Termination 1B, during rapid sea-level rise, a time when sediment supply should have decreased to submarine fan complexes. Weber et al. [1997] attributed this increase in sedimentation to increased runoff associated with intensified monsoon activity in northern India.

It has also been demonstrated that for the continental margin of the Fly River region, Gulf of Papua, New Guinea, significant volumes of terrigenous sediments continued to be supplied to the Coral Sea Basin during much of the sea-level rise (18 to 11 ka) of the last deglaciation [Harris et al., 1996]. This result is a paradox because terrigenous sediment supply should have essentially ceased with backfilling of the river system during the last transgression [Harris et al., 1996]. However, in the Coral Sea Basin, terrigenous sediment deposition continued until 11 ka, after which it ceased and biogenic sediments dominated [Gardner, 1970]. Harris et al. [1996] attempted to explain this sequence of events by invoking a mechanism of coast-parallel sediment transport to a region of narrow shelf where sediment was then funneled into the deep basin. However, this process would still require major sediment supply by the Fly River to the inner continental shelf, an unlikely process during the rapid sea-level rise. This paradox can perhaps be explained by invoking the upper continental slope as the source of terrigenous sediments during the deglacial episode, if it experienced increased instability associated with methane hydrate dissociation.

Likewise, on the upper continental slope of the Gulf of Mexico (150 to 1450 m) the Brazos-Trinity turbidite system displays evidence of maximum downslope sediment transport during the last glacial termination (~11.5 ka) compared to other intervals of the last glacial-interglacial cycle [Badalini et al., 2000].

Furthermore, the record suggests episodic down-slope sediment transport linked to the Bond Cycles, with major reactivation of the turbidite system at the beginning of each Bond Cycle. This reactivation is represented by debris flows deposited through the entire system and increased rates of down-slope sediment transport [Badalini et al., 2000].

The patterns in the various regions studied have generally been interpreted to reflect the effects of sea-level change or earthquakes on upper slope instability— explanations, which seem inadequate. We suggest, instead, that enhanced activation of mass wasting and the turbidite systems resulted, in part, from methane hydrate instability. This hypothesis is supported by (1) the timing of increased turbidite deposition during glacial terminations; and (2) evidence from the Madiera Abyssal Plain turbidite sequence for multiple contemporaneous sediment sources during deposition of individual turbidites [Rothwell et al., 1992]. These occur in both proximal and distal turbidite settings. Rothwell et al. [1992] suggested that it is unlikely that these turbidites originated from single enormous slides but rather from a series of retrogressive slides. Because of this, they suggest that large seismic shocks may have contemporaneously triggered a number of separate slides. Because seismogenic slope failures would be distributed randomly in time, this hypothesis does not explain the episodic emplacement of turbidites during climate change at marine isotope stage transitions. A methane hydrate trigger appears consistent with both contemporaneous triggering of multiple turbidity currents and their activation at times of climate change. Earthquakes may act as a final triggering mechanism once methane hydrates have become potentially unstable.

Turbidite deposition within the ocean can occur at any time, but as documented above, it clearly was episodic during the late Quaternary and appears to have been at a maximum during climate transitions, rather than during peak glacial and interglacial episodes. In previous studies, most discussion of turbidite triggering by methane hydrate dissociation has been associated with maximum lowstands of sea level when hydrostatic pressure was lowest [Paull et al., 1996]. Overall, this does not seem to be supported by our compilation of the data, although individual events have been dated to the LGM. For example, slope failure from hydrate dissociation has been attributed to generation of a megaturbidite layer of 8 to 10 m thickness in the Balearic Basin, western Mediterranean during the LGM at ~22 ka when sea level was at its lowest [Rothwell et al., 1998].

Pockmarks

Pockmarks also represent an apparently widespread manifestation of sea-floor CH_4 release during the latest Quaternary. These features are circular to ellipsoidal shallow depressions on the sea floor, often intimately associated with areas of methane hydrate occurrence [Hovland and Judd, 1988]. A pockmark is likely formed as a result of a violent burst of released gas when hydraulic connection is made between the gas reservoir and ocean [Hovland and Judd, 1988; Uchupi et al., 1996]. As gas pressure builds up in a shallow porous layer beneath an impermeable cohesive sealing layer, excess pore pressure is released by eruption through fractures. Sediments within the pockmark fail by fluidization within the rising gas plume, become entrained and dispersed by bottom currents.

Pockmarks are now widely reported as a result of extensive side-scan sonar surveys of upper continental slopes such as off the Big Sur Coast of California [Paull et al., 2002]. Pockmark fields are widespread on the upper continental slope and often consist of hundreds to thousands of individual pockmarks [Hovland and Judd, 1988; Paull et al., 2002]. For example, a large pockmark field covering more than 1000 km^2 in the Point Arena Basin, northern California, between 1 and 1.5 km water depth, exhibits a density of >3 per km^2 in some areas [McAdoo et al., 2000]. Individual pockmarks are ~0.5 km wide and tens of meters deep. This pockmark field does not seem to be directly associated with slumps, although numerous slumps occur in the general area. The landslides near the pockmark field may be caused by sediment weakening because of gas/fluid venting. The pockmark field off Big Sur, California (560 km^2 area between 900 and 1200 m water depth) [Paull et al., 2002] exhibits 1500 pockmarks, each between 130 to 260 m in diameter and ~8 to 12 m deep.

Timing of Pockmark Activity

The large areas covered by pockmarks suggest an upward flux of CH_4 from an extensive sub-bottom CH_4 source, probably derived from dissociating hydrates. Earthquakes have been suggested to trigger increased activity in pockmark fields [Hasiotis et al., 1996]. Knowledge about the age of pockmark formation and activity from CH_4 release should be of value in determining potential significance, if any, to climate change. The age of pockmark formation can potentially be determined by dating sediments within the structures.

Little is known about the age of formation of most pockmark fields, but evidence exists that their activity is episodic. Most pockmark fields exhibit little to no present day activity. For example, inactive pockmarks in the large field off Big Sur, California may have formed earlier than 45 ka [Paull et al., 2002] although most pockmarks appear fresh in side-scan sonar images. Long et al. [1998] described large (300 to 500 m), deep (10 to 30 m) pockmarks (and associated inferred dormant hydrate mounds) in the Barents Sea. They suggested, from various lines of geological evidence, that these formed during the last deglaciation when sea level was 80 m lower than present. Extensive pockmark fields located on the upper continental slope of northern California (Eel River Basin) are mostly inactive even though they are at water depths (400 to 600 m) within the zone of potential methane hydrate instability [Yun et al., 1999]. Currently, this field both lacks significant gas in associated subsurface sediments and gas plumes in the water column. These data suggest a history of episodic and catastrophic venting of gas during the latest Quaternary [Yun et al., 1999]. Although the age of the pockmark formation is unknown, excellent morphologic preservation suggests relative youth within the latest Quaternary. These observations suggest that the main process that forms pockmark fields is essentially inactive at present, although Hovland and Judd [1988] did observe activity at numerous pockmarks they studied.

We suggest that pockmarks formed as a result of active CH_4 discharge through dissociation of methane hydrates during the last deglaciation. Stabilization of methane hydrates and associated reduction in CH_4 release followed during the Holocene, in conjunction with sea-level rise and cessation of bottom-water temperature oscillations. The existence of extensive fields of pockmarks on the continental slope may represent evidence for past major methane hydrate dissociation on upper continental slopes, although seismic activity may have played a role. It is possible that many pockmark fields formed during the glacial-interglacial transition (see "Future Tests of the Hypothesis" chapter), contributing significantly to the transfer of CH_4 from the methane hydrate reservoir to the ocean/atmosphere system, with possible climate implications.

Clay Deposits, Heinrich Events and Nepheloid Layers

Evidence also exists that widespread fine-grained sediment layers in late Quaternary sediment sequences of the Atlantic Ocean were deposited from nepheloid layers produced as a result of major sediment disturbance on upper continental slopes and potentially within the depth zone of methane hydrate instability. This process appears to be associated with the formation of a gray layer at low latitudes and Heinrich Layers at high northern latitudes in the Atlantic.

Atlantic Gray Layer. A distinct gray clay layer deposited during the interval of the last deglaciation has been described over wide areas of the western equatorial Atlantic and in some locations off the northwest coast of Africa [Ruddiman, 1997; Richardson, 1974; Damuth, 1977; Broecker et al., 1993; Hemming et al., 1998] (Figure 26). This layer is distinctive in that it represents deposition of terrigenous clays in otherwise deep-sea biogenic sediments of latest Quaternary sequences. Detailed studies of the clay layer [Hemming et al., 1998; Ruddiman, 1997] have shown it originated from an increase in clay flux because of downslope sediment redistribution, compatible with a process of mass wasting from different sediment sources on the continental margin. The clay layer appears to have been deposited between ~16 and 14 ^{14}C kyr B.P. (19 to 16.7 ka) (Figure 26). This is within the interval of early, moderate deglaciation and rise in sea level beginning at ~18 ka [Fairbanks, 1989].

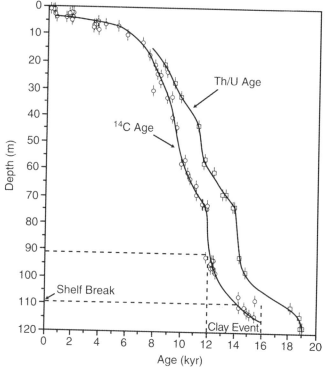

Figure 26. Age of clay layer deposited between 16 and 12 ka in the western tropical Atlantic Ocean [Hemming et al., 1998] with respect to sea-level records from Barbados [Bard et al., 1990]. Deposition of the clay layer occurred during early deglaciation and associated early rapid rise in sea level and during transgression over the continental shelf break.

The origin of this clay layer is controversial, although it clearly reflects a major episode of siliciclastic sediment transfer from upper continental slopes of the equatorial Atlantic into the deep sea. Hemming et al. [1998] suggested two hypotheses for the origin of the clay layer: (1) it represents evidence for major rainfall increase in South America during deglaciation, and a resulting peak in terrigenous sediment transfer to the deep sea; or (2) it resulted from mass wasting associated with the beginning of sea-level rise across the shelf. Although not entirely satisfied with either hypothesis, they tended to favor the second, because of the layer's distribution over such wide areas including the West African margin. However, Lin et al. [1997] recognized a negative planktonic foraminiferal $\delta^{18}O$ event associated with the gray layer (~16.7 ka) in the near-coastal Cariaco Basin, northern Venezuela. They interpreted the negative $\delta^{18}O$ values as reflecting decreased surface ocean salinity, and hence favored the first hypothesis. Nevertheless, although continental precipitation on South America likely increased dramatically during the last glacial termination [Peterson et al., 2000], it seems unlikely that clay transportation by surface currents would have been sufficiently large to form such thick clay deposits in deep and remote parts of the equatorial Atlantic Ocean, especially since Amazon and Oronoco River runoff tends to be constrained within northward-flowing coastal currents and to plumes with limited extension into the open ocean. This would seem to limit the possibility of major sediment transfer to broad areas of the open Atlantic Ocean.

Late Quaternary terrigenous sediments from Ceara Rise in the western tropical Atlantic Ocean are derived from the Amazon margin of the equatorial Atlantic. Harris and Mix [1999] have documented a remarkable suite of peaks in iron oxide and oxyhydroxides (goethite and hematite) associated with glacial-interglacial transitions during the last million years. The most recent of these correlates with an iron-rich layer described throughout the western tropical Atlantic at the last glacial termination and generally attributed to *in situ* diagenesis [McGeary and Damuth, 1973]. Harris and Mix [1999] considered the layers too thick (~25 cm) to have resulted from diagenesis. They instead attributed the layers to increased precipitation in the Amazon Basin that caused increased erosion of chemically weathered lowland sediments relative to physically weathered highland (Andean) sediments. Although this is possible, once again it seems unlikely that such widely distributed iron-rich sediments in the western tropical Atlantic could have been deposited directly from tropical rivers. Furthermore, the deposition of these layers occurred during rapid transgressions associated with deglaciation when sediment entrainment began to occur within the continental watershed.

We instead suggest that the clay layer and iron-rich layers represent episodes of siliciclastic sediment deposition resulting from failure of upper continental slope sediments during the last deglaciation. Furthermore, we suggest that the age, distribution, and character of the clay layer (Figure 26) is consistent with the hypothesis of major increase in slope instability related to methane hydrate dissociation during the last deglaciation. This hypothesis seems to be supported by the age of the most recent episode of massive slumping of the Amazon Fan [Maslin et al., 1998]. The magnitude of such massive slumping should have caused entrainment of clay in intense benthic nepheloid layers (Plate 5) [Kumar and Embley, 1977] at that time. Thus, clays would have been widely transported to the deep-sea basins, distant from continental margins. Clay entrainment in nepheloid layers would have resulted from turbulence associated with slump generation, turbidity currents and other mass sediment flows, and dynamic CH_4 release from the upper continental slope, itself causing sediment input into the water column (Plate 5). Geochemical and mineralogical evidence from the clay layer suggests multiple sediment sources along the margin of South America [Hemming et al., 1998]. We suggest that the age and distribution of the clay layer and its multiple sources are consistent with this hypothesis, providing an additional suggestion of major slope disturbance resulting from methane hydrate instability during the last deglaciation.

Heinrich Layers. Major disturbance of high-latitude continental slope sediments of the Northwest Atlantic also seems to have occurred early during the last deglaciation, as recorded by four mineralogically-distinct layers of "brick red sandy mud" deposited within a so-called Heinrich Layer. Heinrich Layers were initially thought to be made up entirely of ice-rafted sediments [Heinrich, 1988] deposited episodically from armadas of icebergs in the North Atlantic during the late Quaternary. However these sediment layers are too thick (up to 20 cm) to have been deposited as a result of ice-rafting alone. Compelling evidence now indicates that Heinrich Layers were also associated with extensive nepheloid flow activity and even turbidity current deposition in the Northwest Atlantic [Andrews and Tedesco, 1992; Hesse and Khodabakhsh, 1998; Piper and Skene, 1998; Bout-Roumazeilles et al., 1999]. High concentrations of fine-grained carbonate in Heinrich Layers were supplied by nepheloid flows detached from the ocean floor or, in some areas, by turbidity currents [Hesse and Khodabakhsh, 1998]. The layers appear to have been deposited at times of major climate instability: extreme cooling followed by abrupt warming. This caused destabilization and ice loss at ice sheet margins over continental shelves and of ice shelves [Siegart et al., 2002]. Associated rapid sea-level rises of 10 to 15 m support such ice sheet melting during the Heinrich Events as recorded by coral terrace formation on Huon Peninsula, Papua New Guinea [Yokoyama et al., 2001]. At these times increased iceberg discharge was associated with nepheloid layer components of fine sediments in a complex of reworked sediments from multiple sources [Hesse and Khodabakhsh, 1998; Piper and Skene, 1998; Bout-Roumazilles et al., 1999]. Their deposition reflects major disruption of continental margin sediments over broad areas providing a range of sediment types and depositional modes [Hesse and Khodabakhsh, 1998]. In mid-latitude cores from the eastern North Atlantic, Heinrich Layers are marked by fine-grained terrigenous sediment often lacking ice rafted detritus [Bard et al., 2000]. As suggested by Bard et al. [2000], the fine-grained sediment component is unlikely to have been transported in icebergs and is more readily explained by deposition from nepheloid layers, as in other areas of the North Atlantic.

Heinrich Layers are generally considered to have resulted from changes in ice-sheet dynamics [Andrews and Tedesco, 1992; MacAyeal, 1993]. However, high-resolution studies indicate that sediment deposition from nepheloid layers precedes deposition from ice-rafting processes [Bout-Roumazilles et al., 1999]. Such a sequence of events is clearly recorded in Heinrich Event 1 on the Iberian margin [Bard et al., 2000]. The two processes do not seem to be closely dependent. This suggests that the nepheloid layers may not have resulted from changes in ice-sheet dynamics, but through some other process. Perhaps the character, timing, and climate associations of the Heinrich and similar, although thicker, layers of the last deglaciation, are a reflection of major upper continental slope instability in the Northwest Atlantic. This instability could also have resulted from methane hydrate dissociation. This hypothesis is supported by several lines of evidence that include:

- Apparent source of sediments from relatively shallow depths on the upper continental slope.
- Association with turbidites derived by downslope sediment reworking from slumping on the upper continental slope [Wang and Hesse, 1996], the same source areas that produced the nepheloid layers.
- Apparent near-synchronism of depositional events over broad areas and yet from different source regions, suggesting destabilization by some common trigger [Piper and Skene, 1998].
- Deposition at times of significant climate instability that occurred during the last deglaciation and the intervals of Heinrich Events that terminate each of the Bond Cycles.

Magnitude of Methane Release from Slope Instability

Major slumps on the upper continental slope likely transferred large volumes of CH_4 from the methane hydrate reservoir to the ocean/atmosphere system. Large slumps can potentially release enormous volumes of CH_4 (~1 Pg) into the ocean/atmosphere system [Nisbet and Piper, 1998]. Furthermore, large slumps activated by methane hydrate instability appear to be widespread. If correct, this would clearly implicate hydrate instability as an important process in climate change [Nisbet, 1992]. For example, the Storegga submarine slump off Norway [Bugge et al., 1988] is associated with unstable hydrate fields. The Storegga slump was enormous, having transported 5600 km^3 of sediment 800 km from the upper continental slope into the Norwegian Sea Basin. This slump could have rapidly released between ~1 and 5 Pg of CH_4 [Nisbet and Piper, 1998]. Instability of methane hydrates in this region has been exaggerated by post-glacial uplift of the margin [Bugge et al., 1988].

The catastrophic CH_4 releases in Santa Barbara Basin recorded by large, negative benthic and planktonic $\delta^{13}C$ excursions (Plates 1 & 6) [Kennett et al., 2000a] may have been associated with submarine slides [Kennett and Sorlien, 1998]. Although these events could be considered relatively small in terms of aerial extent, they have the potential for significant CH_4 release. Conservative calculations accounting for the magnitude and duration of the isotopic excursion suggest 6.4 Tg yr[-1] was expelled into basin waters, which if transported to the atmosphere, would be equal to ~1.3% of the modern annual flux. At this rate, one brief event may have released ~18% of the average CH_4 increase associated with interstadials [Kennett et al., 2000a]. If methane was efficiently transported to surface waters and the atmosphere, then the total volume of gas released by these events was likely greater than calculated because the amount that escaped into the atmosphere was not recorded by the $\delta^{13}C$ of foraminifera.

10

Role of Methane in Quaternary Climate Change

Atmospheric Methane Effect on Warming

Methane is a potent greenhouse gas that is 23 times more effective at warming than an equivalent volume of CO_2 over a 100-yr time frame, and an astonishingly 62 times more powerful greenhouse gas over a shorter, 20-yr time span [Ehhalt et al., 2001]. The presence of 1.5 ppmv of atmospheric CH_4 would cause globally averaged surface temperature to be 1.3°C higher than without CH_4 [Donner and Ramanathan, 1980]. Doubling of atmospheric CH_4 is generally considered to cause an average global surface temperature rise of ~1°C [Leggett, 1990]. If rapid, such a rise would be significant for global warming because of the promotion of positive feedbacks [Crowley and North, 1991]. Increases in CH_4 directly enhance the trapping of terrestrial infrared radiation, but also perturb tropospheric chemistry, produce O_3 (another greenhouse gas) in the upper troposphere, reduce OH concentrations, and increase tropospheric water vapor [Hanson and Hanson, 1996]. Methane increases also occur in the stratosphere, increasing stratospheric water vapor, an even more powerful greenhouse gas [Cicerone and Oremland, 1988]. Estimates of climate forcing between the years 1850 and 2000 [Hansen et al., 2000] indicate that CH_4 has had a larger influence on modern global warming (0.7 watts m^{-2}) than has generally been recognized, fully half that of the forcing by CO_2 (1.4 watts m^{-2}).

The role of increased water vapor is crucial in assessing the impact of atmospheric CH_4 increases. Even if relative humidity remains fairly constant as CH_4 and consequently temperature rise, an absolute increase in the amount of atmospheric water vapor occurs, which in turn causes further temperature amplification of perhaps 50% [Cicerone and Oremland, 1988].

Methane has a short atmospheric residence time (~20 yr) [Leggett,1990], being rapidly oxidized to CO_2 and H_2O. Thus, sustained climate warming resulting from increased CH_4 requires a continuous supply. Sustained supplies must be

Methane Hydrates in Quaternary Climate Change
© 2003 by the American Geophysical Union
10.1029/054SP11

provided by the major potential sources including methane hydrates, continental wetlands and/or permafrost [Haq, 1995]. Positive feedback processes should have played a major role towards sustained CH_4 input into the atmosphere during global warming.

We suggest that significant atmospheric CH_4 increases can force rapid global warming. However, such warming must have been reinforced by several other processes acting as positive feedbacks. These would have included increases in other greenhouse gases. The feedbacks would have occurred rapidly because much of the initial change occurs within the atmosphere and near the ocean surface (including the melting of sea ice at high latitudes). This is supported by a modeling experiment involving major (4000 Tg) CH_4 release to the atmosphere which was alone insufficient to cause radiative forcing of a magnitude to induce the magnitude of the recorded deglacial warming. However, the model was able to simulate deglacial warming when additional feedbacks were included, specifically combinations of CH_4, CO_2, and heat transport [Thorpe et al., 1998].

Reinforcement from other Greenhouse Gases

Additional reinforcement of the CH_4-initiated global warming would also be provided by CO_2 degassing from a warmer ocean, and increases in atmospheric N_2O which, like CH_4, exhibits rapid variation in close association with rapid temperature rises recorded in the Greenland ice core [Flückiger et al., 1999]. Apart from water vapor, N_2O is the third most important greenhouse gas after CO_2 and CH_4. Like CH_4, N_2O is believed to have an important terrestrial source in the tropics, although unlike CH_4, it is produced in wet forest soils and dry savannas rather than wetlands. N_2O also has an important oceanic source in upwelling regions overlying zones of denitrification [Naqvi et al., 2000].

Significant changes in atmospheric N_2O occur in response to climate oscillations suggesting important changes in N_2O production. The general pattern of N_2O variations recorded in the GRIP ice core is similar to that for CH_4, with lower concentrations during cold episodes and higher concentrations during warm episodes [Flückiger et al., 1999]. Like CH_4, N_2O seems to be coupled to Greenland temperature and therefore to the rapid, late Quaternary climate changes in the Northern Hemisphere. However, important differences in atmospheric concentration of the two gases are apparent in the ice core records. First, the sawtooth pattern characteristic of the CH_4 record is absent in the N_2O record. More pronounced differences occur at the last glacial termination. Both records exhibit a significant increase at the onset of the Bølling. However, during the Younger Dryas, N_2O decreased at a slower rate than CH_4 even when the longer atmospheric lifetime of N_2O is considered. Additionally, N_2O, unlike CH_4, did not reach a usual cold phase level during this cool climate event. Although N_2O sources are as unconstrained as they are for CH_4 and have a similarly intimate relationship with climate change, it is apparent that the sources of the two greenhouse gases have a differential response to climate change [Flückiger et al., 1999].

Water Vapor. Greenhouse gas forcing of rapid climate change likely requires a major role for water vapor. Methane and CO_2 alone do not exhibit sufficiently large increases to force the large, rapid temperature shifts of the late Quaternary. It is likely that changes in the atmospheric inventory of water vapor played a key role [Broecker, 1997a, 1997b]. Planetary warming also stimulates the hydrologic cycle, increasing atmospheric water vapor, reinforcing effects caused by increased CH_4 and CO_2. Water vapor is an effective and major greenhouse gas that depends exponentially on temperature; its doubling can lead to a $1°C$ increase in surface temperature [Oltmans and Hofmann, 1995]. Thus, it can exert significant leverage over climate change and has rapid response times [Pierrehumbert, 1999]. Increased SST would have further amplified and propagated warming via the temperature-water vapor feedback by increasing partial pressure of water vapor. Therefore, warming of the Earth itself stimulates additional greenhouse effects. For example, warming of the ocean's upper layer decreases the solubility of CO_2 and increases atmospheric CO_2, providing greenhouse feedback [Stocker and Schmittner, 1997]. Furthermore, increased atmospheric CH_4 leads to increased water vapor concentrations in the stratosphere by increasing CH_4 oxidation [Oltmans and Hoffmann, 1995; Hanson and Hanson, 1996]. Thus, a number of positive feedback mechanisms exist for contributing to the large, rapid climate warmings of the late Quaternary. The tropics would have played a central role in such warming because this is the dominant region of atmospheric water vapor production.

Rapid increases in inferred precipitation in northern South America were clearly in phase with the stadial-interstadial cycles, as recorded in Amazon margin (Figure 9) [Arz et al., 1998] and Cariaco Basin sequences (tropical Atlantic) (Figure 8) [Peterson et al., 2000]. Likewise, in southeast Asia, because of increased east Asian monsoon strength, interstadials were warmer and wetter as indicated by lower inferred sea-surface salinity in the South China Sea [Wang et al., 1999a, b] and $\delta^{18}O$ changes recorded in speleothems in China [Y. Wang et al., 2001]. Associated warming over broad oceanic areas would have increased atmospheric water vapor. The rapid onset of interstadials appears closely matched with equally rapid rises in inferred precipitation and atmospheric water vapor as the tropics and subtropics expanded [Sachs and Lehman, 1999; Hendy and Kennett, 1999, 2000; Kennett et al., 2000a; Y. Wang et al., 2001]. Surface waters of the North Pacific gyre warmed significantly during deglaciation [Lee and Slowey, 1999] supporting earlier evidence of such change based on snowline estimates on Hawaii [Porter, 1979]. This provided greater water vapor to the atmosphere, with consequent reinforcement of global warming [Lee and Slowey, 1999].

Tropical regions must play a key role in coupling the climate of the Northern and Southern Hemispheres. Furthermore, the tropics are effective in communicating climate change across longitudes because tropical climate is zonally relatively uniform [Pierrehumbert, 1999]. Therefore, it is not surprising that climate change appears to be in phase between northern high latitude regions (Greenland-North Atlantic) [Dansgaard et al., 1993; Bond et al., 1993], the mid-northern latitudes [Sachs and Lehman, 1999; Behl and Kennett, 1996; Hendy and Kennett, 1999, 2000], the tropical Atlantic [Hughen et al., 1996; 2000; Behling et al., 1999; Peterson et al., 2000], and likely through changing monsoon activity in the Arabian Sea [Schulz et al., 1998] and Asia [Wang and Oba, 1998; Y. Wang et al., 2001].

Increasing monsoon strength in association with deglacial warming appears to have been gradual and regionally complex. In northeastern Brazil, increased monsoon strength and associated precipitation occurred during the last glacial termination between 15.5 and 11.8 ^{14}C kyr B.P. (Figure 9) [Behling et al., 1999]. Oxygen isotopic variations recorded in late Quaternary speleothems in China (32°30'N, 119°10'E) indicate an increase in inferred summer monsoon strength beginning at ~16 ka. This was followed by a rapid major increase at the onset of the B-Å (14,645 ± 60 yr B.P.), and apparently synchronous with the rapid warming at the onset of the B-Å in Greenland [Y. Wang et al., 2001]. Records in southeast Asia (South China Sea) suggest a more gradual increase in monsoon strength during the latest Quaternary. In this region the summer monsoon may have strengthened ~3.8 kyr earlier than the first meltwater pulse at Termination 1A [Wang et al., 1999a, b] and continued to steadily strengthen into the early Holocene [Wang et al., 1999a]. The differences in these Asian records may be a reflection of their different resolutions.

The transition from low to high precipitation may have been slow during the glacial-interglacial transition in western tropical South America based on a lacustrine record in the Peruvian Andes [Seltzer et al., 2000]. Peak monsoons and associated precipitation occurred several thousand years after the glacial termination. Models indicate that this occurred between 9 and 8 ka [Short and Mengel, 1986], several thousand years following solstice-perihelion alignment, and as a direct thermal response to a peak in precession. This produced the warmest summers and greatest land-sea temperature contrasts in the tropics. The timing of this peak is approximately coeval with evidence for early Holocene high lake levels and recharged aquifers in north Africa [Street and Grove, 1979; Petit-Maire et al., 1991; Hoelzmann et al., 2000; Abell and Hoelzmann, 2000].

Within the tropics, hydrologic change was out of phase between the northern and southern tropics during the last deglaciation [Thompson et al., 1995, 1998; Seltzer et al., 2000]. Regional differences occur in monsoon strength and its effects on atmospheric water vapor. No uniform response seems to have occurred in the hydrologic cycle between the hemispheres to trigger millennial-scale climate variability. The rapid rises in CH_4 during the last glacial termination were not always in phase with inferred precipitation increases associated with developing monsoon strength. Nevertheless, as discussed earlier, broad areas of the tropics did undergo changes in precipitation in phase with millennial-scale climate variability, thus serving as a vitally important feedback mechanism.

Methane and Quaternary Climate Change: Question of Lead/Lags

The remarkably close association between changes in CH_4 and climate during the late Quaternary suggest a common origin. However, it remains unclear if the changes were synchronous or if small, but significant, lead/lags occurred between the rapid rises in CH_4 and temperature. Solving this remains vital for testing the *Clathrate Gun Hypothesis*. The problem of phasing is difficult to resolve conclusively because of the age difference between ice and gas deposition in ice cores. In the Greenland Ice Sheet at Summit, transfer of atmospheric gases from the surface to the firn-ice transition takes ~12 yr under present climate conditions [Schwander et al., 1993]. The difference between age of ice and mean age of the occluded air is ~210 yr [Schwander et al., 1993]. Gas in the trapped bubbles of ice does not have a discrete age, but rather an age distribution due to diffusion [Schwander, 1996]. Thus, temporal relations between CH_4 and ice in the past have to be modeled using certain assumptions including gas diffusion rates through the firn, effects of thermal diffusion and gravitational settling on composition, and the time required to completely close the gas bubbles [Severinghaus et al., 1998; Severinghaus and Brook, 1999]. Temporal correlation in most studies is thus indirect and temporal resolution insufficiently refined to determine relations between CH_4 and climate change on the necessary ~decadal time scale [Haq, 1999] (see "Future Tests of the Hypothesis" section).

The resolution of the problem may ultimately come from analyses of ice core records of sufficiently high-resolution using approaches that can accurately deconvolve the ice-age/gas-age differences. Such experiments have explored temporal relations between the rapid rise in CH_4 and climate warming during Terminations 1B [Severinghaus et al., 1998] and 1A [Severinghaus and Brook, 1999]. These experiments employed anomalies in $\delta^{15}N$ of N_2 as gas phase markers of warming. Uncertainties in timing because of ice-age/gas-age differences were considered eliminated by Severinghaus et al. [1998] and Severinghaus and Brook [1999] using an indication of warming recorded in the trapped gases based on the $\delta^{15}N$ ratio of N_2. Within the firn, ^{15}N diffuses to colder and ^{14}N to warmer levels, creating a gradient in $^{15}N/^{14}N$ ratio determined by the temperature gradient. The nitrogen isotopic composition of gas in ice bubbles depends on the difference in temperature between the surface and at depth at the time of bubble closure. Thus, surface temperature records are established [Taylor, 1999].

Use of this proxy confirms the close temporal relationship between temperature and CH_4 increases at Terminations 1A and 1B, but suggests that the beginning of increase in CH_4 lagged the beginning of temperature rise by 0 to 30 yr (Termination IB) and 20 to 30 yr (Termination 1A). Furthermore, the CH_4

increased more slowly (over 50 to 150 yr) than the rapid temperature rises which occurred within only a few decades [Brook et al., 2000]. During Termination IB, almost all of the inferred atmospheric CH_4 increase occurred after most of the rapid warming had occurred [Severinghaus et al., 1998; Alley and Clark, 1999]. The results of these experiments led Severinghaus et al. [1998] and Severinghaus and Brook [1999] to reject a causal role for CH_4 in climate change. They suggested instead that the CH_4 rise was probably produced by changes in the low-latitude hydrologic cycle and that the dominant source of increasing atmospheric CH_4 was a slightly delayed response in tropical wetland development. This supported earlier conclusions [Brook et al., 1996] that CH_4 increase was a minor contributor to warming during interstadials, but conflicts with the conclusions of Petit et al. [1999] of a major role for greenhouse gas forcing of Quaternary climate change, including CH_4. The results of these experiments, if correct, do not exclude methane hydrates as an important source of the CH_4 increases at glacial and stadial terminations [Brook et al., 1996, 2000]. They would, however, exclude methane hydrates as an important driver of late Quaternary climate change.

Nevertheless, the climate/CH_4 age model of Severinghaus et al. [1998] and Severinghaus and Brook [1999] includes assumptions based on $^{15}N/^{14}N$ fractionation related to temperature gradients in the firn and the uptake of $^{15}N/^{14}N$ ratios and CH_4 in gas bubbles. The conclusions of Severinghaus et al. [1998] and Severinghaus and Brook [1999] seem to lie within the error of the method used. Because of this and other reasons including the following, we have adopted a working hypothesis that major increases in CH_4 and rapid warming were dynamically linked and were essentially synchronous. We find synchroneity compelling because of the great Greenhouse Warming Potential of CH_4 on very short time scales [Ehhalt et al., 2001] that would be unrepresented in the temperature records if CH_4 rise truly lagged warming. Thus we suggest the following:

- A literal interpretation of the model results of Severinghaus et al. [1998] indicates that CH_4 increases almost completely followed most of the rapid warming, and therefore would have been a result of rather than a cause of it. Instead, we follow the interpretation of Petit et al. [1999] that CH_4 (and other greenhouse gases, especially CO_2 [Cuffey and Vimeux, 2001]) did contribute as a positive feedback to rapid warmings at the glacial terminations. Considering methane's large Global Warming Potential on decadal to centennial time scales [Ehhalt et al., 2001], it is unlikely that near doubling of atmospheric CH_4 concentrations at Termination 1A, for example, had negligible effect on global temperatures.

- For reasons discussed in the previous section involving the inability of wetlands to rapidly respond to climate change, continental wetlands are unlikely sources for the rapid rises in CH_4 at the glacial and stadial terminations.

Blunier [2000] considered early deglaciation of Antarctica following the LGM and preceding the last glacial termination as a concern for the *Clathrate Gun Hypothesis*. This early Antarctic deglaciation appears to be matched by early warming and deglaciation prior to Termination 1A (onset of Bølling) over broad areas of the Northern Hemisphere (see [Lagerklint and Wright, 1999] and references therein) including Santa Barbara Basin [Hendy et al., 2002]. This early deglaciation and associated warming caused ~20 m of sea-level rise following the LGM [Hanebuth et al., 2000] and was more gradual and limited compared with the abrupt rise in temperature associated with CH_4 increases at Termination 1A. It is these dramatic changes that the *Clathrate Gun Hypothesis* attempts to explain.

A further objection raised by Blunier [2000] is that the simultaneous CO_2 rises at the glacial terminations were larger than the increases in CH_4, implying greater importance. As discussed earlier, the associated rise in atmospheric CO_2 and of inferred water vapor associated with changes in continental precipitation [e.g., Peterson et al., 2000], represent positive feedbacks needed to amplify warmings initiated by CH_4 releases from methane hydrates. Thus, the *Clathrate Gun Hypothesis* provides a feasible mechanism for sudden release of CH_4 into the atmosphere, triggering greenhouse warming. No known similar mechanism exists for rapid major CO_2 storage and release; although increase in rapid atmospheric CO_2 may be explained by the cumulative effect of a number of processes associated with rapid global warming [Sigman and Boyle, 2000].

11

Role of Methane Hydrates in Quaternary Climate Change

Glacial Terminations

The speed, magnitude, and timing of the glacial terminations during the late Quaternary ice age are consistent with a process involving major atmospheric CH_4 emission from destabilized methane hydrates. The *Clathrate Gun Hypothesis* predicts that the largest exchange of CH_4 from the hydrate reservoir to the ocean/atmosphere system would occur at glacial terminations for several reasons: (1) low sea levels would have decreased pressure on the hydrate reservoir and increased its potential instability creating a hypercritical state for some hydrates [Buffett & Zatsepina, 2000; Buffett, 2000]; (2) expansion of cold intermediate waters during glacial maxima would have caused growth of hydrates on shallower regions of upper continental slopes; and (3) the several thousand year interval during glacial maxima marked by relatively stable, cold deep-sea waters would have caused growth of hydrates especially in areas where continuous upward diffusion of CH_4 adds to the volume of hydrate [Hyndman et al., 2001]. Such conditions should have occurred at shallow depths within the sediment column. At this stage in the glacial cycle, conditions were most favorable for the largest volume of gas to be released by methane hydrate dissociation upon warming of intermediate waters [Kennett et al., 2000a; Hendy and Kennett, submitted]. Consistent with this explanation is the repeated association of the coldest, most glaciated interval, followed by the warmest with distinct CH_4 overshoots observed at some terminations. Highest CH_4 values tend to occur early in interglacial and interstadial episodes which is consistent with this hypothesis.

A critical component of the *Clathrate Gun Hypothesis* is the effect of bottom-water temperature increases at the depths of potential methane hydrate instability (~400 to 1000 m) (Figure 20B). Changes in sea level alone were not sufficient

Methane Hydrates in Quaternary Climate Change
© 2003 by the American Geophysical Union
10.1029/054SP12

or rapid enough to cause major instability. Pressure decrease related to a 100 m drop in sea level is equal to only about one degree or less of bottom-water warming in the zone of potential methane hydrate instability (Figure 20B). Yet, as described earlier, strong evidence exists for significant temperature increase (~3°C) of upper intermediate waters on continental margins during deglaciation, sufficient to destabilize methane hydrates.

Millennial-Scale Climate Variability

The remarkable millennial-scale climate variability during the latest Quaternary, including abrupt global warmings, was driven by processes yet unexplained. The behavior of these short-duration events gives insight into the processes involved in the larger orbital-scale deglaciations. Any hypothesis proposed to explain this climate history must be consistent in explaining the record at all time scales through the late Quaternary. The following major characteristics of millennial-scale climate behavior during the last glacial cycle appear consistent with the *Clathrate Gun Hypothesis*:

- The remarkably rapid warmings at the onset of interstadials.
- Significant climate instability during these warmings [Sachs and Lehman, 1999].
- Close association and similar relative magnitude of temperature and atmospheric CH_4 variation.
- Largest climate variability during MIS 3 when the methane hydrate reservoir was particularly unstable because of low sea-level and intermediate-water temperature oscillations.
- Strong tropical involvement in stadial-interstadial cycles including large changes in temperature and continental precipitation.
- Largest warmings at the onset of interstadials when methane hydrate dissociation and hence atmospheric CH_4 emission would have been greatest.
- Brief, conspicuous temperature overshoots at the onset of many interstadials recorded in polar ice cores and marine sediment suggestive of dynamic temperature increases.
- Brief, conspicuous CH_4 overshoots at the onset of several interstadials suggesting dynamic CH_4 release from methane hydrates.
- Sawtooth pattern of temperature and CH_4 during interstadials because of rapid increases and slower decreases.
- Clustering of stadial-interstadial cycles into groups (Bond Cycles) that initiate with the warmest interstadial and subsequently decrease in magnitude. Bond Cycle terminations are marked by the most extended cool episode (Heinrich Event) that immediately precedes the next major

warming marking the onset of the next Bond Cycle. Thus, the coldest episodes are juxtaposed with the warmest, much like that of the 100-kyr cycle. The resulting sawtooth pattern is thus consistent with cycles that begin with major hydrate dissociation following extended cool periods followed by interstadial episodes of decreasing magnitude. This pattern is consistent with initial dissociation of the most accessible, near surface methane hydrates followed by the gradual accumulation of methane hydrates as colder intermediate waters become progressively more dominant and warm intermediate waters during interstadials become weaker and less extensive. The magnitude and duration of interstadials are also likely modulated by orbital forcing (tilt and precession), that affects the extent, temperature and duration of thermohaline circulation changes in a non-linear relationship. Thus MIS 5 and early MIS 3 warm episodes were of longer duration and generally warmer than those of MIS 4 and later MIS 3.

• Millennial-scale climate change is synchronous between the high northern latitudes, middle latitudes, and tropics and between terrestrial and marine systems. This is consistent with a greenhouse forcing mechanism and teleconnections via the atmosphere.

• Rapid (decades to centuries) interstadial terminations. These intervals are matched in Santa Barbara Basin by sudden, rapid bottom-water cooling, reflecting a sudden switching to cool intermediate waters. This evidence is consistent with a sudden decrease in CH_4 emissions because of the methane hydrate reservoir stabilization. The brevity of interstadials (500 to 2000 yr) is thus linked to the duration of warm intermediate-water episodes that largely control methane hydrate reservoir stability. Initial warming of upper intermediate waters is amplified into abrupt, major warming marking the onset of interstadial and interglacial episodes by release of CH_4 from dissociating methane hydrates. This forcing of temperature change in upper intermediate waters operates at a more rapid pace than orbital cycles. The nature of this suborbital pacemaker remains largely unknown and represents one of the most vexing problems in earth sciences.

A problem of potential importance relates to the observation that interstadials 19 and 20 are not associated with large rises in CH_4 (Figure 4) [Brook et al., 1999], the only such episodes that exhibit this behavior. These interstadials are of large magnitude and duration and should exhibit associated CH_4 rises just as with other interstadials. Instead, CH_4 rises were small. Thus, at face value, the absence of CH_4 peaks during interstadials 19 and 20 would suggest a near-absence of wetlands, or absence of any emissions from methane hydrates. If the rapid warmings that mark the onset of interstadials 19 and 20 occurred in absence of CH_4 rises, this would seem to negate the hypothesis that CH_4 variations contributed to temperature oscillations during the late Quaternary [Brook et al., 1996], unless in the unlikely case they were forced by some other mechanism. This remains a mystery. We suspect that absence of major CH_4 peaks are caused by other factors, such as lack of signal preservation or brevity of events as yet undetected.

Millennial-scale (sub-Milankovitch) climate variability has been recorded in marine sediment sequences older than one million years, and thus prior to the development of the 100-kyr regime [Raymo et al., 1998; Ortiz et al., 1999]. Insufficiently high-resolution records of this age have made it difficult to resolve the detailed character of this variability and to determine if behavior was similar to that during the latest Quaternary. However, it is unlikely that the behavior will ultimately be shown to be similar because of differences in sensitivity of the climate system to feedbacks. The *Clathrate Gun Hypothesis* predicts that greater stability of sea level and thermohaline circulation before 800 ka would result in less variable intermediate-water temperature and hence greater stability and lesser involvement of the methane hydrate reservoir in climate change.

Origin of 100-kyr Cycle

A fundamental change occurred in the behavior of the Earth's climate system at ~900 ka with the initiation of the 100-kyr cycle. The magnitude of the marine oxygen isotope oscillations increases (Figure 27) over a ~300 kyr interval, beginning shortly after 1 Ma and evolving over several glacial cycles towards a distinct 100-kyr dominated periodicity by ~800 ka. The increase in magnitude is conspicuous mainly because of the $\delta^{18}O$ enrichment during the glacial phase of each cycle reflecting Northern Hemisphere ice sheet expansion [Imbrie et al., 1993]. Since then, the system has continued to evolve more slowly, but does exhibit a gradual shift towards higher amplitude climate and ice volume change (Figure 27). Ice sheets during glacial episodes seem to have become slightly larger, and interglacial episodes warmer (Figure 27).

Two major features of Quaternary climate behavior remain unexplained: (1) the cause of the late Quaternary dominance of the 100-kyr periodicity of eccentricity, by far the weakest insolation forcing [Imbrie et al., 1993; Raymo, 1997]; and, (2) why this dominance over the stronger precession and obliquity cycles began relatively suddenly after ~900 ka. The essence of the 100-kyr problem is the difficulty in finding a physically plausible linear mechanism that would amplify the system's response to a small insolation forcing in a narrow band of periods near 100 kyr [Imbrie et al., 1993]. Muller and MacDonald [1995, 1997] attempted to explain this through variations in inclination of the Earth's orbit, but this hypothesis does not seem feasible for reasons outlined by Berger [1999].

Sharp, distinct glacial terminations did not exist before the 100-kyr regime. Therefore, the feedbacks to the system were not as effective in promoting large, rapid shifts in climate. After ~900 ka, some critical threshold of the system was

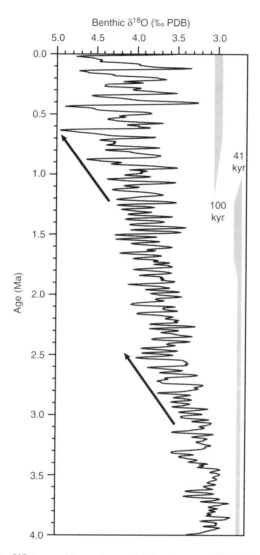

Figure 27. Benthic $\delta^{18}O$ record from Ocean Drilling Program Site 846, eastern equatorial Pacific from the middle Pliocene (4 Ma) to the Present [Shackleton et al., 1995; Mix et al., 1995]. This record reflects late Neogene evolution of climate variability modes and magnitude as the global cryosphere expanded. Arrows indicate major global steps in late Neogene glacial development; between 3 and 2.5 Ma representing early growth of Northern Hemisphere ice sheets; and between 1.2 and 0.7 Ma reflecting the transition (shown also by vertical bars at right) from orbital (Milankovitch) variations dominated by 41-kyr cycles (obliquity; "41-kyr world") to 100-kyr cycles (eccentricity; "100-kyr world"). Note also increased amplitude of 41-kyr cycles before the onset of 100-kyr cycles, which in turn are amplified during the early part of this new state.

passed so that the relatively small amplitude input of Milankovitch oscillation was amplified by feedbacks internal to the earth system. Several internal processes have been suggested to explain the magnitude of the warming at the glacial terminations at 100-kyr periodicity. These include instability of Northern Hemisphere ice sheets resulting from critical maximum growth that lead to rapid ice sheet surges [MacAyeal, 1993], and a threshold change in the sensitivity of thermohaline circulation during the mid-Quaternary in response to either critical ice sheet growth [Raymo, 1997] or tectonic changes at high north Atlantic latitudes [Denton et al., 1999]. Denton et al. [1999] suggested that the development of the 100-kyr cycle resulted from evolution of modern thermohaline circulation such that the system became easier to switch off and harder to switch on. Other hypotheses have been summarized by Hinnov [2000] and include changes in frequency resonance among the orbital parameters [Liu, 1992, 1998; Rial, 1999]. Liu [1998] argued that a 100-kyr resonance phenomenon developed between the Earth's obliquity and eccentricity during the late Quaternary. However the evolution of the 100-kyr periodicity in the late Quaternary generated by these models has required the assistance of an additional external agent such as decreasing atmospheric CO_2 [Hinnov, 2000].

Shackleton [2000] suggested that the 100-kyr signal did not arise from the long time constant of ice volume response, but rather through deep-sea temperature change, which is highly coherent with the 100-kyr orbital band of the Antarctic (Vostok ice core) CO_2 signal. Contrary to previous investigations, Shackleton [2000] found global ice volume to lag changes in orbital eccentricity. Atmospheric CO_2, Vostok air temperature and deep-water temperatures are in phase with orbital eccentricity at the 100-kyr period. The effect of orbital eccentricity probably enters the paleoclimatic record through influence on concentrations of atmospheric CO_2 [Shackleton, 2000]. This suggests that CO_2 has direct and immediate control on deep-water temperature via atmospheric temperature at high latitudes. Recent support for this comes from δD temperature corrections that reveal significant covariation of temperature and CO_2 in the Vostok ice core providing additional compelling evidence that CO_2 is an important forcing factor for climate [Cuffey and Vimeux, 2001]. This conclusion is consistent with the *Clathrate Gun Hypothesis* because CH_4 is largely in phase with CO_2, particularly during the rapid glacial terminations, which are the intervals of the 100-kyr cycle exhibiting the largest magnitude climate shifts.

We suggest that the evolution of the 100-kyr regime resulted from critical development of intermediate-water behavior as it affected the thermal regime and hence instability of the methane hydrate reservoir. The large climate feedbacks that rapidly propelled the Earth from glacial to interglacial state during major terminations did not exist before ~900 ka. The *Clathrate Gun Hypothesis* invokes major CH_4 emissions at these terminations because of hydrate instability. The eccentricity cycle is emphasized because its insolation minimum, when reinforced by minima in the precession and obliquity cycles, produced the longest, coldest glacial intervals with lowest sea level at the end of each 100-kyr cycle. These glacial maxima were colder than during previous glacial episodes and allowed maximum recharge of the methane hydrate reservoir, providing greater volumes of accessible hydrates sensitive to orbitally driven intermediate-water warming. The major destabilizing mechanisms of the methane hydrate reservoir were significantly reduced before the onset of the 100-kyr world for three main reasons: (1) Ice sheets, and hence sea-level oscillations were smaller; (2) the magnitude of glacial-interglacial climate change, and hence ocean temperature oscillations were smaller; and (3) thermohaline circulation had not evolved to its modern state, including the amplitude of glacial-interglacial temperature variation of intermediate waters. The origin of the 100-kyr cycle was linked to the critical growth of Northern Hemisphere ice-sheets and related climate regime. Once ice sheets grew beyond a critical threshold, conditions were created that promoted methane hydrate accumulation necessary for subsequent massive atmospheric CH_4 emissions. In turn, CH_4 release promoted rapid warming reinforced by feedback mechanisms that forced the dramatic climate warming at glacial terminations.

Origin of Sawtooth Pattern

Late Quaternary climate change was marked by rapid warmings followed by gradual coolings at orbital and millennial time scales. This general pattern of coolest episode followed by warmest episode resulted in the pervasive sawtooth pattern of climate behavior. This is observed in: (1) the basic glacial-interglacial 100-kyr cycle; (2) interstadials of decreasing robustness (briefer and of lower amplitude) through the 100-kyr cycle; (3) the Bond Cycles; (4) the end of each Bond Cycle marked by an especially cold episode, representing the beginning of Heinrich Events; (5) the succession of Bond Cycles through MIS 3 generally composed of interstadials of decreasing magnitude and duration; (6) individual interstadials; and, (7) temperature overshoots at the onset of some glacial and stadial terminations which exaggerate the sawtooth pattern.

It appears that the magnitude of interstadials was modulated by orbital forcing. Interstadials were of longer duration and larger magnitude when Northern Hemisphere insolation was greater through combination of tilt, eccentricity and precession. The strength of the feedbacks reinforcing each interstadial was modulated by its position within the 100-kyr cycle. The pattern of diminishing interstadial strength is consistent with initial depletion of the most easily dissociated methane hydrate deposits followed by the gradual accumulation of hydrates in association with progressively increasing dominance of colder intermediate waters. The final coldest stadials reset the process to begin a new cycle beginning with maximum warmth. The magnitude of interstadials also generally decreased throughout the duration of the last 100-kyr cycle as the Earth became progressively cooler, peaking with maximum cooling at the LGM. A general relationship exists between the magnitude of coolings and that of the succeeding warmings. Thus colder intervals and/or coolings of longer duration were followed by especially warm episodes. Glacial terminations are the most extreme examples of this behavior. Thus, the Bølling interstadial of interglacial magnitude followed the LGM, and the Heinrich Events that terminate each Bond Cycle were associated with a jump into a warm interstadial state.

The same relationships also seem to have been exhibited by MIS 11 and 12, the climate behavior of which has remained unexplained. MIS 12 was one of the coldest glacial maxima during the late Quaternary. It was followed immediately by one of the warmest interglacials (MIS 11) and hence this glacial termination was one of the most dramatic. Yet MIS 11 occurred during an interval of low orbital eccentricity and thus of low precessional amplitude [Imbrie et al., 1993]. As in other parts of the late Quaternary, the magnitude of the warming at the termination appears to have been set by the extent of cooling that immediately preceded it.

Any hypothesis attempting to explain late Quaternary climate change must account for the pervasive sawtooth pattern, and none proposed thus far have satisfactorily accomplished this. However, the sawtooth climate behavior appears consistent with the *Clathrate Gun Hypothesis*: rapid warmings resulted from CH_4 discharge into the atmosphere from the largest accumulation of methane hydrates. This provided important new energy input into the climate system to accelerate initiation of the warm episodes. Massive CH_4 release would be especially effective at rapidly shifting climate into an interglacial or interstadial state because of its very large Greenhouse Warming Potential on short time scales [Ehhalt et al., 2001]. The largest warmings occurred at the onset of each warm episode. The more readily accessible methane hydrate is depleted such that succeeding CH_4 discharges into the atmosphere become increasingly diminished resulting in successively smaller interstadials. Recharging of the hydrate reservoir intensified during the extended coolings at the end of each 100-kyr glacial episode and the Bond Cycles. This sets the climate system for larger successive warming episodes, at glacial terminations and Bond Cycle initiations. Strong support comes from the co-variation of temperature and atmospheric CH_4 oscillations [Petit et al., 1999; Brook et al., 1999]. Both exhibit the sawtooth pattern on orbital and millennial time scales. The magnitude of change in both is remarkably similar and provides compelling evidence that they were dynamically related.

12

Future Tests of the Hypothesis

In this contribution, we have integrated a wide range of data from ice cores and marine and continental sedimentary and paleontological records that, collectively formulated as the Clathrate Gun Hypothesis, make a compelling explanation of many aspects of late Quaternary climate behavior. Nearly all of the relevant data have come from studies conducted during the last decade (as reflected in the reference list) indicating the youthfulness of this new synthesis. Given this, there are many fruitful components of the hypothesis to test. Some of these are suggested below, arranged by the component tested. Although given the breadth of the hypothesis, it is likely that a number of rigorous tests will be required to confirm or negate the various elements that make up the hypothesis.

Atmospheric Methane Record of Ice Cores

- Refine models and tests of the effects of gas diffusion, gravitational separation, and the gas-age/ice-age difference on the phasing and rate-of-change of CH_4 relative to temperature in ice cores during episodes of rapid environmental change.
- Investigate the influence of post-incorporation microbial and chemical processes on preserving or modifying atmospheric gas concentrations in the ice, especially that of CH_4.
- Determine validity and frequency of brief CH_4 "overshoots" associated with rapid CH_4 increases in ice cores through high temporal resolution investigations.

Methane Hydrates in Quaternary Climate Change
© 2003 by the American Geophysical Union
10.1029/054SP13

Source of Atmospheric Methane

- Determine the stable isotopic composition of C and H from CH_4 in ice cores as an indication of its source—methane hydrate or terrestrial wetlands.
- Determine ^{14}C activity of CH_4 in ice cores from samples of known age to differentiate contemporaneous wetland sources from predicted older CH_4 derived from methane hydrates.

Wetland History

- Refine timing of wetland initiation and evolution in several critical areas including the large modern ecosystems in southeast Asia.
- Determine wetland extent and variability during the millennial-scale climate oscillations of the last glacial episode.

Marine Methane Hydrate Stability

- Determine spatial and temporal variability of inferred major CH_4 emissions from methane hydrates during the late Quaternary and compare them with the ice core CH_4 records.
- Strengthen understanding of proxies for past CH_4 emissions to the water column from methane hydrates including $\delta^{13}C$, organic biomarkers, and faunal assemblages.
- Better determine the spatial and temporal history of mass sediment wasting from continental slopes for comparison with predicted intervals of methane hydrate instability during the late Quaternary.
- Determine if evidence exists, such as pockmark fields, for widespread methane hydrate distribution at or near the ocean floor during predicted intervals of methane hydrate instability during the late Quaternary.

Paleoceanography/Paleoclimatology

- Improve global coverage of records of rapid, millennial-scale climate oscillations during the late Quaternary to determine potential inter-hemispheric climate synchroneity that would have resulted from greenhouse gas forcing.
- Better understand spatial and temporal variability of upper intermediate water thermohaline circulation especially as it affects bottom water temperature within the depth zone of potential methane hydrate instability (~400 to 1000 m) on continental slopes.
- Determine whether anomalous ^{14}C reservoir ages documented for the deglacial episode and other times during the late Quaternary were

affected by release of ^{14}C-depleted CH_4 from methane hydrates rather than by changes in ocean circulation.

- Investigate earlier episodes of climate behavior like that of the late Quaternary ("100-kyr world") to test the potential role of methane hydrates as a cause of this climate variability.

Modeling

- Expand models of CH_4 atmospheric chemistry to accommodate the full range of variation in fluxes, sources, and sinks likely encountered during the late Quaternary, especially with respect to changes in atmospheric CH_4 residence time related to inferred rapid, major releases from methane hydrates.
- Refine climate models by incorporating additional feedbacks associated with rapid atmospheric CH_4 increases.
- Refine models of late Quaternary changes in the inter-polar CH_4 gradient to incorporate the potential contributions of methane hydrates as globally distributed at various times on the continental margins.
- Model the variation in size and distribution of the methane hydrate reservoir through the late Quaternary as constrained by changes in intermediate water temperature, sea level, and ice coverage.

13

Conclusions

- Unlike the prevailing interpretation that continental wetlands were the principal source for the rapid atmospheric CH_4 increases during the late Quaternary, we suggest a marine sedimentary methane hydrate source.
- Negligible wetland ecosystems existed during the last glacial episode as a result of global aridity, low sea level, incised, well-flushed river systems and low water tables.
- Wetland ecosystems were insufficiently developed during the last glacial episode to account for the rapid atmospheric CH_4 increases during glacial and stadial terminations recorded in polar ice cores.
- The large, modern wetland ecosystems (peatlands, tropical floodplains, and coastal wetlands) developed almost exclusively during the Holocene, well after the rapid atmospheric CH_4 increases during the last glacial termination.
- Widespread evidence exists that the upper continental slope methane hydrate reservoir, rather than being stable during the late Quaternary as often suggested, exhibited episodic and structured instability.
- Methane hydrate instability was modulated by changing sources and temperature of upper intermediate water. Warmer intermediate waters during interglacials and interstadials fostered dissociation of methane hydrates while cooler intermediate waters during glacials and stadials replenished and expanded the reservoir. Sea level also played a role by increasing the vulnerability of methane hydrates to dissociation during glacial terminations when sea level was especially low. As a consequence, the largest atmospheric CH_4 increases occurred during glacial terminations. The process that forced changes in upper intermediate waters, especially on millennial time scales, remains to be determined.

Methane Hydrates in Quaternary Climate Change
© 2003 by the American Geophysical Union
10.1029/054SP14

- We suggest that CH_4 outgassing from the methane hydrate reservoir played a key role in late Quaternary climate behavior, an integrative hypothesis that appears consistent with a wide range of paleoclimatic and marine geological data, and which we term the *Clathrate Gun Hypothesis*.

- Late Quaternary methane hydrate reservoir instability resulted in upper continental slope instability marked by slumping, nepheloid layers, pockmarks, turbidite deposition and increased sediment transport to deep-sea basins. Consistent with the *Clathrate Gun Hypothesis*, mounting evidence suggests that upper continental slope instability was especially enhanced during glacial terminations rather than glacial maxima.

- We suggest that instability of the methane hydrate reservoir has played a critical role in late Quaternary climate change. The methane hydrate reservoir is effectively a stored energy source, accumulating during cold intervals and released during warmings. We suggest that reservoir instability produced sufficient emissions of methane into the atmosphere to amplify initial warmings and cause abrupt climatic shifts.

- Atmospheric CH_4 emissions from methane hydrates effectively "jump-started" the rapid warmings that marked the late Quaternary promulgating other feedback mechanisms including other greenhouse gases (H_2O and CO_2), necessary for enhancing the warming.

- The *Clathrate Gun Hypothesis* potentially explains aspects of late Quaternary climate behavior that have long been enigmatic including the rapidity of warmings, large initial warming followed by gradual cooling, sawtooth pattern of climate behavior, Bond Cycles, and 100-kyr cycle. The hypothesis is also consistent in explaining the close relationship between atmospheric CH_4 and climate variability.

- The late Quaternary, an interval marked by the dominance of the 100-kyr cycle, has been an unusual time in the Earth's climate history marked by unusually strong and frequent climate changes. Climate feedbacks have rarely been so large. We suggest that this resulted from particularly large changes in stability of the methane hydrate reservoir. This instability stems from relatively large and frequent oscillations in temperature of upper intermediate waters on upper continental margins. Marked instability of upper intermediate waters is a feature that resulted from evolution to critical state of the cryosphere in the Northern Hemisphere during the late Quaternary. The unusual climate behavior of the late Quaternary appears to have been matched by unusual behavior of the methane hydrate reservoir.

References

Abell, P. I., and P. Hoelzmann, Holocene paleoclimates in northwestern Sudan: Stable isotope studies on molluscs, *Global Planet. Change*, *26*, 1-12, 2000.

Abell, P. I., and I. Plug, The Pleistocene/Holocene transition in South Africa: Evidence for the Younger Dryas event, *Global Planet. Change*, *26*, 173-179, 2000.

Adams, J. M., H. Faure, L. Faure-Denard, J. M. McGlade, and F. I. Woodward, Increases in the terrestrial carbon storage from the Last Glacial Maximum to the present, *Nature*, *348*, 711-714, 1990.

Alcalde, J. A., and J. J. Kulemeyer, The Holocene in the south-eastern region of the Province Jujuy, North-West Argentina, *Quat. Int.*, *57/58*, 113-116, 1999.

Allen, J. R. M., U. Brandt, A. Brauer, H.-W. Hubberten, B. Huntley, J. Keller, M. Kraml, A. Mackensen, J. Mingram, J. F. W. Negendank, N. R. Nowaczyk, H. Oberhänsli, W. A. Watts, S. Wulf, and B. Zolitschka, Rapid environmental changes in southern Europe during the last glacial period, *Nature*, *400*, 740-743, 1999.

Alley, R. B., The Younger Dryas cold interval as viewed from central Greenland, *Quat. Sci. Rev.*, *19*, 213-226, 2000.

Alley, R. B., and P. U. Clark, The deglaciation of the northern hemisphere: A global perspective, *Annu. Rev. Earth Planet. Sci.*, *27*, 149-182, 1999.

Alley, R. B., D. A. Meese, C. A. Shuman, A. J. Gow, K. C. Taylor, P. M. Grootes, J. C. W. White, M. Ram, E. D. Waddington, P. A. Mayewski, and G. A. Zielinski, Abrupt increase in Greenland snow accumulation at the end of the Younger Dryas event, *Nature*, *362*, 527-529, 1993.

Alley, R. B., R. C. Finkel, K. Nishiizumi, S. Anandakrishnan, C. A. Shuman, G. R. Mershon, G. A. Zielinski, and P. A. Mayewski, Changes in continental and sea-salt atmospheric loadings in central Greenland during the most recent deglaciation, *J. Glaciol.*, *41*, 503-514, 1995.

Alley, R. B., P. A. Mayewski, T. Sowers, M. Stuiver, K. C. Taylor, and P. U. Clark, Holocene climate instability: A prominent, widespread event 8200 years ago, *Geology*, *25*, 483-486, 1997.

Alley, R. B., S. Anandakrishnan, and P. Jung, Stochastic resonance in the North Atlantic, *Paleoceanography*, *16*, 190-198, 2001.

Aloisi, G., C. Pierre, J. M. Rouchy, J. P. Foucher, and J. Woodside, Methane-related authigenic carbonates on eastern Mediterranean Sea mud volcanoes and their possible relation to gas hydrate destabilization, *Earth Planet. Sci. Lett.*, *184*, 321-338, 2000.

Alperin, M. J., and W. S. Reeburgh, Geochemical observations supporting anaerobic methane oxidation, in *Microbial Growth on C-1 Compounds*, edited by R. Crawford and R. Hanson, pp. 282-289, American Society for Microbiology, Washington, D.C., 1984.

Altabet, M. A., M. J. Higginson, and D. W. Murray, The effect of millennial-scale changes in Arabian Sea denitrification on atmospheric CO_2, *Nature*, *415*, 159-162, 2002.

Alve, E., and J. M. Bernhard, Vertical migratory response of benthic foraminifera to controlled oxygen concentrations in an experimental mesocosm, *Mar. Ecol. Prog. Ser.*, *116*, 137-151, 1995.

Anderson, J. A. R., The tropical peat swamps of western Malesia, in *Mires: Swamp, Bog, Fen, and Moor: General Studies*, edited by A. J. P. Gore and D. W. Goodall, *Ecosyst. World*, *4A*, pp. 181-196, Elsevier, New York, 1983.

Andrews, J. T., and K. Tedesco, Detrital carbonate-rich sediments, northwestern Labrador Sea: Implications for ice-sheet dynamics and iceberg rafting (Heinrich) events in the north Atlantic, *Geology*, *20*, 1087-1090, 1992.

Angstadt, D. M., J. A. Austin, Jr., and R. T. Buffler, Deep sea erosional unconformity in the southeastern Gulf of Mexico, *Geology*, *11*, 215-218, 1983.

Arz, H. W., J. Pätzold, and G. Wefer, Correlated millennial-scale changes in surface hydrography and terrigenous sediment yield inferred from last-glacial marine deposits off northeastern Brazil, *Quat. Res.*, *50*, 157-166, 1998.

Aselmann, I., and P. J. Crutzen, Global distribution of natural freshwater wetlands and rice paddies, their net primary productivity, seasonality and possible methane emissions, *J. Atmos. Chem.*, *8*, 307-358, 1989.

Badalini, G., B. C. Kneller, and C. D. Winker, Architecture and processes in the Late Pleistocene Brazos-Trinity turbidite system, Gulf of Mexico continental slope, in *Deep-Water Reservoirs of the World*, edited by P. Weimer, pp. 16-34, Gulf Coast Section SEPM Foundation, Houston, 2000.

Bains, S., R. M. Corfield, and R. D. Norris, Mechanisms of climate warming at the end of the Paleocene, *Science*, *285*, 724-727, 1999.

Bains, S., R. D. Norris, R. M. Cornfield, and K. L. Faul, Termination of global warmth at the Palaeocene/Eocene boundary through productivity feedback, *Nature*, *407*, 171-174, 2000.

Bak, P., *How Nature Works: The Science of Self-Organized Criticality*, 212 pp., Copernicus, New York, 1996.

Baker, P. A., G. O. Seltzer, S. C. Fritz, R. B. Dunbar, M. J. Grove, P. M. Tapia, S. L. Cross, H. D. Rowe, and J. P. Broda, The history of South American tropical precipitation for the past 25,000 years, *Science*, *291*, 640-643, 2001.

Baraza, J., G. Ercilla, and C. H. Nelson, Potential geologic hazards on the eastern Gulf of Cadiz slope (SW Spain), *Mar. Geol.*, *155*, 191-215, 1999.

Bard, E., M. Arnold, J. Duprat, J. Moyes, and J.-C. Duplessy, Reconstruction of the last deglaciation: Deconvolved records of $\delta^{18}O$ profiles, micropaleontological variations and accelerator mass spectrometric ^{14}C dating, *Clim. Dyn.*, *1*, 101-112, 1987.

Bard, E., B. Hamelin, R. G. Fairbanks, and A. Zindler, Calibration of the ^{14}C timescale over the past 30,000 years using mass spectrometric U-Th ages from Barbados corals, *Nature*, *345*, 405-410, 1990.

Bard, E., F. Rostek, and C. Sonzogni, Interhemispheric synchrony of the last deglaciation inferred from alkenone palaeothermometry, *Nature*, *385*, 707-710, 1997.

Bard, E., F. Rostek, J.-L. Turon, and S. Gendreau, Hydrological Impact of Heinrich Events in subtropical northeast Atlantic, *Science*, *289*, 1321-1324, 2000.

Barker, P. A., F. A. Street-Perrott, M. J. Leng, P. B. Greenwood, D. L. Swain, R. A. Perrott, R. J. Telford, and K. J. Ficken, A 14,000-year oxygen isotope record from diatom silica in two alpine lakes on Mt. Kenya, *Science*, *292*, 2307-2310, 2001.

Barnes, R. O., and E. D. Goldberg, Methane production and consumption in anoxic marine sediments, *Geology*, *4*, 297-300, 1976.

Bartlett, K. B., and R. C. Harriss, Review and assessment of methane emissions from the wetlands, *Chemosphere*, *26*, 261-320, 1993.

Bartlett, K. B., P. B. Crill, J. A. Bonassi, J. E. Richey, and R. C. Harriss, Methane Flux from the Amazon River floodplain: Emissions during rising water, *J. Geophys. Res.*, *95*, 16773-16788, 1990.

Baulin, V. V., and N. S. Danilova, Dynamics of a late Quaternary permafrost in Siberia, in *Late Quaternary Environments of the Soviet Union*, edited by A. A. Velichko, H. E. Wright and C. W. Barnosky, pp. 69-77, University of Minnesota Press, Minneapolis, 1984.

Beck, J. W., D. A. Richards, R. L. Edwards, B. W. Silverman, P. L. Smart, D. J. Donahue, S. Hererra-Osterheld, G. S. Burr, L. Calsoyas, A. J. T. Jull, and D. Biddulph, Extremely large variations of atmospheric ^{14}C concentration during the last glacial period, *Science*, *292*, 2453-2458, 2001.

Behl, R. J., and J. P. Kennett, Brief interstadial events in the Santa Barbara Basin, NE Pacific, during the past 60 kyr, *Nature*, *379*, 243-246, 1996.

Behling, H., Late Quaternary vegetation, climate, and fire history from the tropical mountain region of Morro de Itapeva, SE Brazil, *Palaeogeogr. Palaeoclimatol. Palaeoecol.*, *129*, 407-422, 1997.

Behling, H., and H. Hooghiemstra, Environmental history of the Colombian savannas of the Llanos Orientales since the Last Glacial Maximum from lake records El Pinal and Carimagua, *J. Paleolimnol.*, *21*, 461-476, 1999.

Behling, H., and M. Lima da Costa, Holocene environmental changes from the Rio Curuá record in the Caxiuanã Region, eastern Amazon Basin, *Quat. Res.*, *53*, 369-377, 2000.

Behling, H., and H. Hooghiemstra, Neotropical savanna environments in space and time: Late Quaternary interhemispheric comparisons, in *Interhemispheric Climate Linkages*, edited by V. Markgraf, pp. 307-323, Academic Press, San Diego, 2001.

Behling, H., J. C. Berrio, and H. Hooghiemstra, Late Quaternary pollen records from the middle Caquetá River basin in central Colombian Amazon, *Palaeogeogr. Palaeoclimatol. Palaeoecol.*, *145*, 193-213, 1999.

Behling, H., H. W. Arz, J. Pätzold, and G. Wefer, Late Quaternary vegetational and climate dynamics in northeastern Brazil, inferences from marine core GeoB 3104-1, *Quat. Sci. Rev.*, *19*, 959-994, 2000.

Bell, P. R., Methane hydrate and the carbon dioxide question, in *Carbon Dioxide Review*, edited by W. C. Clark, pp. 401-406, Oxford University Press, New York, 1982.

Bender, M. L., T. Sowers, M. L. Dickson, J. Orchardo, P. Grootes, P. A. Mayewski, and D. A. Meese, Climate connection between Greenland and Antarctica during the last 100,000 years, *Nature*, *372*, 663-666, 1994.

Bender, M. L., B. Malaize, J. Orchardo, T. Sowers, and J. Jouzel, High precision correlations of Greenland and Antarctic ice core records over the last 100 kyr, in *Mechanisms of Global Climate Change at Millennial Time Scales*, edited by P. U. Clark, R. S. Webb and L. D. Keigwin, *Geophys. Monogr. Ser.*, *112*, pp. 149-164, AGU, Washington D.C., 1999.

Bennett, K. D., S. G. Haberle, and S. H. Lumley, The last glacial-Holocene transition in southern Chile, *Science*, *290*, 325-328, 2000.

Benson, L. V., D. Currey, Y. Lao, and S. Hostetler, Lake-size variations in the Lahontan and

Bonneville basins between 13,000 and 9000 ^{14}C yr B.P., *Palaeogeogr. Palaeoclimatol. Palaeoecol.*, *95*, 19-32, 1992.

Berger, W. H., The 100-kyr ice-age cycle: Internal oscillation or inclinational forcing?, *Int. J. Earth Sci.*, *88*, 305-316, 1999.

Berger, W. H., and E. Jansen, Mid-Pleistocene climate shift: The Nansen connection, in *The Polar Oceans and Their Role in Shaping the Global Environment: The Nansen Centennial Volume*, edited by O. M. Johanessen, R. D. Muench and J. E. Overland, *Geophys. Monogr. Ser.*, *85*, pp. 295-311, AGU, Washington D.C., 1994.

Berger, W. H., C. B. Lange, and A. Weinheimer, Silica depletion of the thermocline in the eastern North Pacific during glacial conditions: Clues from Ocean Drilling Program Site 893, Santa Barbara Basin, California, *Geology*, *25*, 619-622, 1997.

Bernhard, J. M., K. R. Buck, and J. P. Barry, Monterey Bay cold-seep biota: Assemblages, abundance and ultrastructure of living foraminifera, *Deep Sea Res., Part I*, *48*, 2233-2249, 2001.

Betancourt, J. L., The Amazon reveals its secrets—partly, *Science*, *290*, 2274-2275, 2000.

Betancourt, J. L., C. Latorre, J. A. Rech, J. Quade, and K. A. Rylander, A 22,000 year record of monsoonal precipitation from northern Chile's Atacama Desert, *Science*, *289*, 1542-1546, 2000.

Blunier, T., "Frozen" methane escapes from the sea floor, *Science*, *288*, 68-69, 2000.

Blunier, T., and E. J. Brook, Timing of millennial-scale climate change in Antarctica and Greenland during the last glacial period, *Science*, *291*, 109-112, 2001.

Blunier, T., J. Chappellaz, J. Schwander, J. M. Barnola, T. Desperts, B. Stauffer, and D. Raynaud, Atmospheric methane record from a Greenland ice core over the last 1,000 years, *Geophys. Res. Lett.*, *20*, 2219-2222, 1993.

Blunier, T., J. Chappellaz, J. Schwander, B. Stauffer, and D. Raynard, Variations in atmospheric methane concentration during the Holocene epoch, *Nature*, *374*, 46-49, 1995.

Blunier, T., J. Chappellaz, J. Schwander, A. Dällenbach, B. Stauffer, T. F. Stocker, D. Raynaud, J. Jouzel, H. B. Clausen, C. Hammer, U., and S. J. Johnsen, Asynchrony of Antarctic and Greenland climate change during the last glacial period, *Nature*, *394*, 739-743, 1998.

Bøe, R., M. Hovland, A. Instanes, L. Rise, and S. Vasshus, Submarine slide scars and mass movements in Karmsundet and Skudenesfjorden, southwestern Norway: Morphology and evolution, *Mar. Geol.*, *167*, 147-165, 2000.

Bohacs, K., and J. Suter, Sequence stratigraphic distribution of Coaly Rocks: Fundamental controls and paralic examples, *AAPG Bull.*, *81*, 1612-1639, 1997.

Bohrmann, G., J. Greinert, E. Suess, and M. Torres, Authigenic carbonates from the Cascadia subduction zone and their relation to gas hydrate stability, *Geology*, *26*, 647-650, 1998.

Boles, J. R., J. F. Clark, I. Leifer, and L. Washburn, Temporal variation in natural methane seep rate due to tides, Coal Oil Point area, California, *J. Geophys. Res.*, *106*, 27077-27086, 2001.

Bond, G. C., The North Atlantic's 1500-year cycle in the South Atlantic, *EOS Trans. AGU*, *81*(48), Fall Meet. Suppl., Abstract OS11E-03, 2000.

Bond, G., H. Heinrich, W. Broecker, L. Labeyrie, J. McManus, J. Andrews, S. Huon, R. Jantschik, S. Clasen, C. Simet, K. Tedesco, M. Klas, G. Bonani, and S. Ivy, Evidence for massive discharges of icebergs into the North Atlantic ocean during the last glacial period, *Nature*, *360*, 245-249, 1992.

Bond, G., W. Broecker, S. Johnsen, J. McManus, L. Labeyrie, J. Jouzel, and G. Bonani, Correlations between climate records from North Atlantic sediments and Greenland ice, *Nature*, *365*, 143-147, 1993.

Bond, G., W. Showers, M. Cheseby, R. Lotti, P. Almasi, P. deMenocal, P. Priore, H. Cullen, I. Hajdas, and G. Bonani, A pervasive millennial-scale cycle in North Atlantic Holocene and glacial climates, *Science*, *278*, 1257-1266, 1997.

Bond, G. C., W. Showers, M. Elliot, M. Evans, R. Lotti, I. Hajdas, G. Bonani, and S. Johnson, The North Atlantic's 1-2 kyr climate rhythm: Relation to Heinrich events, Dansgaard/Oeschger cycles and the Little Ice Age, in *Mechanisms of Global Climate Change at Millennial Time Scales*, edited by P. U. Clark, R. S. Webb and L. D. Keigwin, *Geophys. Monogr. Ser.*, *112*, pp. 35-58, AGU, Washington D.C., 1999.

Bond, G., B. Kromer, J. Beer, R. Muscheler, M. N. Evans, W. Showers, S. Hoffmann, R. Lotti-Bond, I. Hajdas, and G. Bonani, Persistent solar influence on North Atlantic climate during the Holocene, *Science*, *294*, 2130-2136, 2001.

Bonnefille, R., and F. Chalié, Pollen-inferred precipitation time-series from equatorial mountains, Africa, the last 40 kyr BP, *Global Planet. Change*, *26*, 25-50, 2000.

Boone, D. R., Ecology of methanogenesis, in *Microbial Production and Consumption of Greenhouse Gases: Methane, Nitrogen Oxides, and Halomethanes*, edited by J. E. Rogers and W. B. Whitman, pp. 57-70, American Society for Microbiology, Washington, D.C., 1991.

Booth, J. S., D. W. O'Leary, P. Popenoe, and W. W. Danforth, U.S. Atlantic continental slope landslides: Their distribution, general attributes, and implications, in *Submarine Landslides: Selective Studies in the U. S. Exclusive Economic Zone*, edited by W. C. Schwab, H. J. Lee and D. C. Twichell, *U.S. Geological Survey Bulletin*, *B2002*, pp. 14-22, 1993.

Borowski, W. S., C. K. Paull, and W. Ussler, III, Marine pore-water sulfate profiles indicate *in situ* methane flux from underlying gas hydrate, *Geology*, *24*, 655-658, 1996.

Borowski, W. S., C. K. Paull, and W. Ussler, III, Global and local variations of interstitial sulfate gradients in deep-water, continental margin sediments: Sensitivity to underlying methane and gas hydrates, *Mar. Geol.*, *159*, 131-154, 1999.

Botch, M. S., and A. Masing, Mire ecosystems in the U.S.S.R., in *Mires: Swamp, Bog, Fen, and Moor: Regional Studies*, edited by A. J. P. Gore and D. W. Goodall, *Ecosyst. World*, *4B*, pp. 95-152, Elsevier, New York, 1983.

Botch, M. S., K. I. Kobak, T. S. Vinson, and T. P. Kolchugina, Carbon pools and accumulation in peatlands of the former Soviet Union, *Global Biogeochem. Cycles*, *9*, 37-46, 1995.

Bout-Roumazeilles, V., E. Cortijo, L. Labeyrie, and P. Debrabant, Clay mineral evidence of nepheloid layer contributions to the Heinrich layers in the northwest Atlantic, *Palaeogeogr. Palaeoclimatol. Palaeoecol.*, *146*, 211-228, 1999.

Bradbury, J. P., B. Leyden, M. Salgado-Labouriau, W. M. Lewis, Jr., C. Schubert, M. W. Binford, D. G. Frey, D. R. Whitehead, and F. H. Weibezahn, Late Quaternary environmental history of Lake Valencia, Venezuela, *Science*, *214*, 1299-1305, 1981.

Bradbury, J. P., M. Grosjean, S. Stine, and F. Sylvestre, Full and late lake records along the PEP 1 transect: Their role in developing interhemispheric paleoclimate interactions, in *Interhemispheric Climate Linkages*, edited by V. Markgraf, pp. 265-291, Academic Press, San Diego, 2001.

Bratton, J. F., Clathrate eustacy: Methane hydrate melting as a mechanism for geologically rapid sea-level fall, *Geology*, *27*, 915-918, 1999.

Bray, F., *The Rice Economies, Technology and Development in Asian Societies*, 254 pp., University of California Press, Los Angeles, 1986.

Brewer, P. G., F. M. Orr, Jr., G. Friederich, K. A. Kvenvolden, D. L. Orange, J. McFarlane, and W. Kirkwood, Deep-ocean field test of methane hydrate formation from a remotely operated vehicle, *Geology*, *25*, 407-410, 1997.

Brewer, P. G., E. T. Peltzer, G. Friederich, and K. A. Kvenvolden, *In situ* ocean experiments on methane hydrate formation and decomposition (abstract), *EOS Trans. AGU*, *80*, Spring Meet. Suppl., 337, 1999.

Brewer, P. G., C. Paull, E. T. Peltzer, W. Ussler, G. Rehder, and G. Friederich, Experimental evidence for the rapid transfer of gas hydrates from the sea floor to the ocean surface, *Geol. Soc. Am. Abstr. Prog.*, *32*, 35, 2000.

Brewer, P. G., C. Paull, E. T. Peltzer, W. Ussler, G. Rehder, and G. Friederich, Experimental evidence for the rapid transfer of gas hydrates from the sea floor to the ocean surface, *Geophys. Res. Letters,* in press.

Broecker, W. S., The salinity history of the northern Atlantic during the last deglaciation, *Paleoceanography*, *5*, 459-467, 1990.

Broecker, W. S., The great ocean conveyor, *Oceanography*, *4*, 79-89, 1991.

Broecker, W. S., Chaotic climate, *Sci. Amer.*, *273*, 62-68, 1995.

Broecker, W. S., Thermohaline circulation, the Achilles Heel of our climate system: Will man-made CO_2 upset the current balance?, *Science*, *278*, 1582-1588, 1997a.

Broecker, W. S., Mountain glaciers: Recorders of atmospheric water vapor content?, *Global Biogeochem. Cycles*, *11*, 589-597, 1997b.

Broecker, W. S., Paleocean circulation during the last deglaciation: A bipolar seesaw?, *Paleoceanography*, *13*, 119-121, 1998.

Broecker, W. S., What if the conveyor were to shut down? Reflections on a possible outcome of the great global experiment, *GSA Today*, *9*, 1-7, 1999.

Broecker, W. S., Abrupt climate change: Causal constraints provided by the paleoclimate record, *Earth Sci. Rev.*, *51*, 137-154, 2000.

Broecker, W. S., and S. Hemming, Climate swings come into focus, *Science*, *294*, 2308-2309, 2001.

Broecker, W. S., D. M. Peteet, and D. Rind, Does the ocean-atmosphere system have more than one mode of operation?, *Nature*, *315*, 21-25, 1985.

Broecker, W. S., G. Bond, and M. Klas, A salt oscillator in the glacial Atlantic? 1. The concept, *Paleoceanography*, *5*, 469-477, 1990.

Broecker, W. S., G. Bond, and J. McManus, Heinrich events: Triggers of ocean circulation changes?, in *Ice in the Climate System*, edited by W. R. Peltier, *NATO ASI Ser.*, *112*, pp. 161-166, Springer-Verlag, Berlin, 1993.

Brook, E. J., T. Sowers, and J. Orchardo, Rapid variations in atmospheric methane concentration during the past 110,000 years, *Science*, *273*, 1087-1091, 1996.

Brook, E. J., S. Harder, J. Severinghaus, and M. Bender, Atmospheric methane and millennial scale climate change, in *Mechanisms of Global Climate Change at Millennial Time Scales*, edited by P. U. Clark, R. S. Webb and L. D. Keigwin, *Geophys. Monogr. Ser.*, *112*, pp. 165-175, AGU, Washington D.C., 1999.

Brook, E. J., S. Harder, J. Severinghaus, E. J. Steig, and C. M. Sucher, On the origin and timing of rapid changes in atmospheric methane during the last glacial period, *Global Biogeochem. Cycles*, *14*, 559-572, 2000.

Brooks, J. M., H. B. Cox, W. R. Bryant, M. C. Kennicutt, II, R. G. Mann, and T. J.

McDonald, Association of gas hydrates and seepage in the Gulf of Mexico, *Org. Geochem.*, *10*, 221-234, 1986.

Brooks, J. M., M. E. Field, and M. C. Kennicutt, II, Observations of gas hydrates in marine sediments, offshore northern California, *Mar. Geol.*, *96*, 103-109, 1991.

Brown, R. J. E., Occurrence of permafrost in Canadian peatlands, paper presented at 3rd International Peatland Congress, Québec, Canada, 1968.

Bubier, J. L., and T. R. Moore, An ecological perspective on methane emissions from the northern wetlands, *Trends Ecol. Evol.*, *9*, 460-464, 1994.

Buffett, B. A., Clathrate hydrates, *Annu. Rev. Earth Planet. Sci.*, *28*, 477-507, 2000.

Buffett, B. A., and O. Y. Zatsepina, Metastability of gas hydrate, *Geophys. Res. Lett.*, *26*, 2981-2984, 1999.

Bugge, T., Submarine slides on the Norwegian continental margin, with special emphasis on the Storegga area, Ph.D. thesis, Norwegian Institute of Technology, Oslo, Norway, 1983.

Bugge, T., R. H. Belderson, and N. H. Kenyon, The Storegga slide, *Philos. Trans. R. Soc. London*, *325*, 357-388, 1988.

Campin, J.-M., T. Fichefet, and J.-C. Duplessy, Problems with using radiocarbon to infer ocean ventilation rates for past and present climates, *Earth Planet. Sci. Lett.*, *165*, 17-24, 1999.

Cane, M. A., and M. Evans, Do the tropics rule?, *Science*, *290*, 1107-1108, 2000.

Cannariato, K. G., and J. P. Kennett, Climatically related millennial-scale fluctuations in strength of California margin oxygen-minimum zone during the past 60 k.y., *Geology*, *27*, 975-978, 1999.

Cannariato, K. G., J. P. Kennett, and R. J. Behl, Biotic response to late Quaternary rapid climate switches in Santa Barbara Basin: Ecological and evolutionary implications, *Geology*, *27*, 63-66, 1999.

Cao, M., K. Gregson, and S. Marshall, Global methane emissions from wetlands and its sensitivity to climate change, *Atm. Environ.*, *32*, 3293-3299, 1998.

Carignano, C. A., Late Pleistocene to recent climate change in Córdoba Province, Argentina: Geomorphological evidence, *Quat. Int.*, *57/58*, 117-134, 1999.

Carpenter, G., Coincident sediment slump/clathrate complexes on the U.S. Atlantic continental slope, *Geo Mar. Lett.*, *1*, 29-32, 1981.

Channell, J. E. T., J. S. Stoner, D. A. Hodell, and C. D. Charles, Geomagnetic paleointensity for the last 100 kyr from the sub-antarctic South Atlantic: A tool for inter-hemispheric correlation, *Earth Planet. Sci. Lett.*, *175*, 145-160, 2000.

Chanton, J. P., J. E. Bauer, P. A. Glaser, D. I. Siegel, C. A. Kelley, S. C. Tyler, E. H. Romanowicz, and A. Lazrus, Radiocarbon evidence for the substrates supporting methane formation within northern Minnesota peatlands, *Geochim. Cosomochim. Acta*, *59*, 3663-3668, 1995.

Chappell, J., Contrasting Holocene sedimentary geologies of lower Daly River, northern Australia, and lower Sepik-Ramu, Papua New Guinea, *Sediment. Geol.*, *83*, 339-358, 1993.

Chappellaz, J., J. M. Barnola, D. Raynaud, Y. S. Korotkevich, and C. Lorius, Ice-core record of atmospheric methane over the past 160,000 years, *Nature*, *345*, 127-131, 1990.

Chappellaz, J., T. Blunier, D. Raynaud, J. M. Barnola, J. Scwander, and B. Stauffer, Synchronous changes in atmospheric CH$_4$ and Greenland climate between 40 and 8 kyr BP, *Nature*, *366*, 443-445, 1993.

Chappellaz, J., T. Blunier, S. Kints, A. Dällenbach, J. M. Barnola, J. Schwander, D.

Raynaud, and B. Stauffer, Changes in the atmospheric CH_4 gradient between Greenland and Antarctica during the Holocene, *J. Geophys. Res.*, *102*, 15987-15997, 1997.

Charles, C. D., J. Lynch-Stieglitz, U. S. Ninnemann, and R. G. Fairbanks, Climate connections between the hemisphere revealed by deep sea sediment core/ice core correlations, *Earth Planet. Sci. Lett.*, *142*, 19-27, 1996.

Charman, D. J., R. Aravena, C. L. Bryant, and D. D. Harkness, Carbon isotopes in peat, DOC, CO_2, and CH_4 in a Holocene peatland on Dartmoor, southwest England, *Geology*, *27*, 539-542, 1999.

Chen, Z. Q., J. F. Donoghue, R. W. Hoenstine, and J. L. Ladner, Late Quaternary history of the northeastern Gulf of Mexico coast, northwest Florida, *Geol. Soc. Am. Abstr. Prog.*, *31*, 423, 1999.

Cicerone, R. J., and R. S. Oremland, Biogeochemical aspects of atmospheric methane, *Global Biogeochem. Cycles*, *2*, 299-327, 1988.

Clapperton, C., *Quaternary Geology and Geomorphology of South America*, 779 pp., Elsevier, Amsterdam, 1993.

Clark, J. F., L. Washburn, J. S. Hornafius, and B. P. Luyendyk, Dissolved hydrocarbon flux from natural marine seeps to the southern California Bight, *J. Geophys. Res.*, *105*, 11509-11522, 2000.

Clark, P. U., and A. C. Mix, Ice sheets by volume, *Nature*, *406*, 689-690, 2000.

Clark, P. U., S. J. Marshall, G. K. C. Clarke, S. W. Hostetler, J. M. Licciardi, and J. T. Teller, Freshwater forcing of abrupt climate change during the last deglaciation, *Science*, *293*, 283-287, 2001.

Clark, P. U., J. X. Mitrovica, G. A. Milne, and M. E. Tamisiea, Sea-level fingerprinting as a direct test for the source of global meltwater pulse 1A, *Science*, *295*, 2438-2441, 2002.

Clark, P. U., N. G. Pisias, T. F. Stocker, and A. J. Weaver, The role of the thermohaline circulation in abrupt climate change, *Nature*, *415*, 863-869, 2002.

Clarke, J. W., P. St. Amand, and M. Matson, Possible cause of plumes from Bennett Island, Soviet Far Arctic, *AAPG Bull.*, *70*, 574, 1986.

Claypool, G., and K. Kvenvolden, Methane and other hydrocarbon gases in marine sediment, *Annu. Rev. Earth Planet. Sci.*, *11*, 299-327, 1983.

Coleman, J. M., and D. B. Prior, Mass wasting on continental margins, *Annu. Rev. Earth Planet. Sci.*, *16*, 101-119, 1988.

Colinvaux, P. A., P. E. De Oliveira, J. E. Moreno, M. C. Miller, and M. B. Bush, A long pollen record from Lowland Amazonia: Forest and cooling in glacial times, *Science*, *274*, 85-88, 1996.

Colinvaux, P. A., P. E. De Oliveira, and M. B. Bush, Amazonian and neotropical plant communities on glacial time-scales: The failure of the aridity and refuge hypotheses, *Quat. Sci. Rev.*, *19*, 141-169, 2000.

Collett, T. S., Natural gas hydrates of the Prudhoe Bay and Kuparuk River area, North Slope, Alaska, *AAPG Bull.*, *77*, 793-812, 1993.

Conrad, R., Control of methane production in terrestrial ecosystems, in *Exchange of Trace Gases Between Terrestrial Ecosystems and the Atmosphere*, edited by M. O. Andreae and D. S. Schimel, pp. 39-58, John Wiley and Sons, New York, 1989.

Crill, P. M., K. B. Bartlett, R. C. Harriss, E. Gorham, E. S. Verry, D. I. Sebacher, L. Madzar, and W. Sanner, Methane flux from Minnesota peatlands, *Global Biogeochem. Cycles*, *2*, 371-384, 1988.

Crill, P. M., R. C. Harriss, and K. B. Bartlett, Methane fluxes from terrestrial wetland environments, in *Microbial Production and Consumption of Greenhouse Gases: Methane, Nitrogen Oxides, and Halomethanes*, edited by J. E. Rogers and W. B. Whitman, pp. 91-109, American Society for Microbiology, Washington, D.C., 1991.

Crowley, T. J., Ice age terrestrial carbon changes revisited, *Global Biogeochem. Cycles*, *9*, 377-389, 1995.

Crowley, T. J., and G. R. North, *Paleoclimatology*, 339 pp., Oxford University Press, New York, 1991.

Cuffey, K. M., and F. Vimeux, Covariation of carbon dioxide and temperature from the Vostok ice core after deuterium-excess correction, *Nature*, *412*, 523-527, 2001.

Cuffey, K. M., G. D. Clow, R. B. Alley, M. Stuiver, E. D. Waddington, and R. W. Saltus, Large Arctic temperature change at the glacial-Holocene transition, *Science*, *270*, 455-458, 1995.

Curry, W. B., and D. W. Oppo, Synchronous, high frequency oscillations in tropical sea surface temperatures and North Atlantic deep water production during the last glacial cycle, *Paleoceanography*, *12*, 1-14, 1997.

Curry, W. B., J. C. Duplessy, L. D. Labeyrie, and N. J. Shackleton, Changes in the distribution of $\delta^{13}C$ of deep water ΣCO_2 between the last glaciation and the Holocene, *Paleoceanography*, *3*, 317-341, 1988.

Curry, W. B., T. M. Marchitto, J. F. McManus, D. W. Oppo, and K. L. Laarkamp, Millennial-scale changes in ventilation of the thermocline, intermediate, and deep waters of the glacial North Atlantic, in *Mechanisms of Global Climate Change at Millennial Time Scales*, edited by P. U. Clark, R. S. Webb and L. D. Keigwin, *Geophys. Monogr. Ser.*, *112*, pp. 59-76, AGU, Washington D.C., 1999.

Cynar, F. J., and A. A. Yayanos, The distribution of methane in the upper waters of the Southern California Bight, *J. Geophys. Res.*, *97*, 11269-11285, 1992.

Dällenbach, A., T. Blunier, J. Flückiger, B. Stauffer, J. Chappellaz, and D. Raynaud, Changes in the atmospheric CH_4 gradient between Greenland and Antarctica during the last glacial and the transition to the Holocene, *Geophys. Res. Lett.*, *27*, 1005-1008, 2000.

Damuth, J. E., Late Quaternary sedimentation in the western equatorial Atlantic, *Geol. Soc. Am. Bull.*, *88*, 695-710, 1977.

Dansgaard, W., S. J. Johnsen, H. B. Clausen, D. Dahl-Jensen, N. Gundestrup, C. U. Hammer, H. Oeschger, North Atlantic climatic oscillations revealed by deep Greenland ice cores, in *Climate Processes and Climate Sensitivity*, edited by J. E. Hansen and T. Takahashi, *Geophys. Monogr. Ser.*, *29*, pp. 288-298, AGU, Washington, D.C., 1984.

Dansgaard, W., S. J. Johnsen, H. B. Clausen, D. Dahl-Jensen, N. S. Gundestrup, C. U. Hammer, C. S. Hvidberg, J. P. Steffensen, A. E. Sveinbjörnsdottir, J. Jouzel, and G. Bond, Evidence for general instability of past climate from a 250-kyr ice-core record, *Nature*, *364*, 218-220, 1993.

deMenocal, P. B., W. F. Ruddiman, and E. M. Pokras, Influences of high- and low- latitude processes on African terrestrial climate: Pleistocene eolian records from equatorial Atlantic Ocean Drilling Program Site 663, *Paleoceanography*, *8*, 209-242, 1993.

deMenocal, P., J. Ortiz, T. Guilderson, and M. Sarnthein, Coherent high- and low-latitude climate variability during the Holocene warm period, *Science*, *288*, 2198-2202, 2000.

Denton, G. H., and W. Karlén, Holocene climatic variations—Their pattern and possible cause, *Quat. Res.*, *3*, 155-205, 1973.

Denton, G. H., and C. H. Hendy, Documentation of an advance on New Zealand's Franz Joseph Glacier at the onset of Younger Dryas time, *Science*, *264*, 1434-1437, 1994.

Denton, G. H., C. J. Heusser, T. V. Lowell, P. I. Moreno, B. G. Andersen, L. E. Heusser, C. Schlüchter, and D. R. Marchant, Interhemispheric linkage of paleoclimate during the last glaciation, *Geogr. Ann.*, *81A*, 107-153, 1999.

Devol, A. H., J. E. Richey, B. R. Forsberg, and L. A. Martinelli, Seasonal dynamics in methane emissions from the Amazon River floodplain to the Troposphere, *J. Geophys. Res.*, *95*, 16417-16426, 1990.

Dickens, G. R., Methane oxidation during the late Paleocene thermal maximum, *Bull. Soc. Geol. Fr.*, *171*, 37-49, 2000.

Dickens, G. R., Sulfate profiles and barium fronts in sediment on the Blake Ridge: Present and past methane fluxes through a large gas hydrate reservoir, *Geochim. Cosomochim. Acta*, 65, 529-543, 2001a.

Dickens, G. R., On the fate of past gas: What happens to methane released from a bacterially mediated gas hydrate capacitor?, *Geochem. Geophys. Geosyst.*, *2*, Paper number 2000GC000131, Jan. 9, 2001b.

Dickens, G. R., Modeling the global carbon cycle with a gas hydrate capacitor: Significance for the latest Paleocene thermal maximum, in *Natural Gas Hydrates: Occurrence, Distribution, and Detection*, edited by C. K. Paull and W. P. Dillon, *Geophys. Monogr. Ser.*, *124*, pp. 19-38, AGU, Washington D.C., 2001c.

Dickens, G. R., and M. S. Quinby-Hunt, Methane hydrate in pore water: A simple theoretical approach for geophysical applications, *J. Geophys. Res.*, *102*, 773-783, 1997.

Dickens, G. R., J. R. O'Neil, D. K. Rea, and R. M. Owen, Dissociation of oceanic methane hydrate as a cause of the carbon isotope excursion at the end of the Paleocene, *Paleoceanography*, *10*, 965-971, 1995.

Dickens, G. R., M. M. Castillo, and J. C. G. Walker, A blast of gas in the latest Paleocene: Simulating first-order effects of massive dissociation of methane hydrate, *Geology*, *25*, 259-262, 1997.

Dickens, G. R., T. J. Bralower, D. J. Thomas, E. Thomas, and J. C. Zachos, High-resolution records of the late Paleocene thermal maximum and circum-Caribbean volcanism: Is there a causal link?: Comment and Reply, *Geology*, *26*, 670–671, 1998.

Dickens, G. R., G. B. Dunbar, M. Page, and A. D. Heap, Massive clastic input to slopes of the Northeast Australian margin during transgression: A new paradigm for sediment behavior in tropical mixed siliciclastic/carbonate systems, paper presented at 7th International Conference on Paleoceanography, Sapporo, Japan, 2001.

Diemont, W. H., and A. Supardi, Accumulation of organic matter and inorganic constituents in a peat dome in Sumatra, Indonesia, paper presented at International Peat Society Symposium on Tropical Peats and Peatland for Development, Yogyakarta, 1987.

Dillon, W. P., Seismic interpretation of gas hydrates in the Blake Ridge area, paper presented at Natural Gas Research and Development Contractors Review Meeting, 1991.

Dillon, W. P., and C. K. Paull, Marine gas hydrates II: Geophysical evidence, in *Natural Gas Hydrates: Properties, Occurrence, and Recovery*, edited by J. L. Cox, pp. 73-90, Butterworth, Boston, 1983.

Dillon, W. P., K. Fehlhaber, M. W. Lee, J. S. Booth, and C. K. Paull, Methane hydrates in sea floor sediments off the Southeastern U.S.: Amounts and implications for climate change, *Geol. Soc. Am. Abstr. Prog.*, *23*, 22, 1991.

Dillon, W. P., J. W. Nealon, M. H. Taylor, M. W. Lee, R. M. Drury, and C. H. Anton, Seafloor collapse and methane venting associated with gas hydrate on the Blake Ridge—Causes and implications to seafloor stability and methane release, in *Natural Gas Hydrates:*

Occurrence, Distribution, and Detection, edited by C. K. Paull and W. P. Dillon, *Geophys. Monogr. Ser.*, *124*, pp. 211-233, AGU, Washington D.C., 2001.

Ditlevsen, P. D., H. Svensmark, and S. Johnsen, Contrasting atmospheric and climate dynamics of the last-glacial and Holocene periods, *Nature*, *379*, 810-812, 1996.

Dokken, T. M., and E. Jansen, Rapid changes in the mechanism of ocean convection during the last glacial period, *Nature*, *401*, 458-461, 1999.

Donner, L., and V. Ramanathan, Methane and nitrous oxide: Their effects on terrestrial climate, *J. Atmos. Sci.*, *37*, 119-124, 1980.

Doose, P. R., and I. R. Kaplan, Biogenic control of gases in marine sediments of Santa Barbara Basin, California, *AAPG Bull.*, *65*, 919-920, 1981.

Doran, P. T., J. C. Priscu, W. B. Lyons, J. E. Walsh, A. G. Fountain, D. M. McKnight, D. L. Moorhead, R. A. Virginia, D. H. Wall, G. D. Clow, C. H. Fritsen, C. P. McKay, A. N. Parsons, Antarctic climate cooling and terrestrial ecosystem response, *Nature*, *415*, 517-520, 2002.

Dowdeswell, J. A., M. A. Maslin, J. T. Andrews, and I. N. McCave, Iceberg production, debris rafting, and the extent and thickness of Heinrich layers (H1, H2) in North Atlantic sediments, *Geology*, *23*, 301-304, 1995.

Duplessy, J.-C., N. J. Shackleton, R. G. Fairbanks, L. D. Labeyrie, D. Oppo, and N. Kallel, Deepwater source variations during the last climatic cycle and their impact on the global deepwater circulation, *Paleoceanography*, *3*, 343-360, 1988.

Ehhalt, D., M. Prather, F. Dentener, R. Derwent, E. Dlugokencky, E. Holland, I. Isaksen, J. Katima, V. Kirchhoff, P. Matson, P. Midgley, and M. Wang, Atmospheric chemistry and greenhouse gases, in *Climate Change 2001: The Scientific Basis: Contribution of Working Group I to the Third Assessment Report of the Intergovernmental Panel on Climate Change*, edited by J. T. Houghton, Y. Ding, D. J. Griggs, and M. Noguer, Cambridge University Press, pp. 239-287, 2001.

Eichhubl, P., H. G. Greene, T. Naehr, and N. Maher, Structural control of fluid flow: Offshore fluid seepage in the Santa Barbara Basin, California, *J. Geochem. Explor.*, *69-70*, 545-549, 2000.

Eichhubl, P., H. G. Greene, and N. Maher, Physiography of an active transpressive margin basin: High-resolution bathymetry of the Santa Barbara Basin, southern California continental borderland, *Mar. Geol.*, *184*, 95-120, 2002.

Elias, S. A., S. K. Short, and R. L. Phillips, Paleoecology of late-glacial peats from the Bering land bridge, Chukchi Sea shelf region, northwestern Alaska, *Quat. Res.*, *38*, 371-378, 1992.

Elias, S. A., S. K. Short, C. H. Nelson, and H. H. Birks, Life and times of the Bering land bridge, *Nature*, *382*, 60-63, 1996.

Elliot, M., L. Labeyrie, G. Bond, E. Cortijo, J.-L. Turon, N. Tisnerat, and J.-C. Duplessy, Millennial-scale iceberg discharges in the Irminger Basin during the last glacial period: Relationship with the Heinrich events and environmental settings, *Paleoceanography*, *13*, 433-446, 1998.

Emery, K. O., and D. Hoggan, Gases in marine sediments, *AAPG Bull.*, *42*, 2174-2188, 1958.

Emery, K. O., R. L. Wigley, A. S. Bartlett, M. Rubin, and E. S. Barghoorn, Freshwater peat on the continental shelf, *Science*, *158*, 1301-1307, 1967.

Emmer, E., and R. C. Thunell, Nitrogen isotope variation in Santa Barbara Basin sediments: Implications for denitrification in the eastern tropical North Pacific during the last 50,000 years, *Paleoceanography*, *15*, 377-387, 2000.

Evans, C. D. R., C. P. Brett, J. W. C. James, and R. Holmes, Shallow seismic reflection profiles from the waters of east and southeast Asia: An interpretation manual and atlas, *BGS Tech. Rep. WC/94/60*, 66-67, 1995.

Evans, D., E. L. King, N. H. Kenyon, C. Brett, and D. Wallis, Evidence for long-term instability in the Storegga Slide region off western Norway, *Mar. Geol.*, *130*, 281-292, 1996.

Eyles, N., Earth's glacial record and its tectonic setting, *Earth Sci. Rev.*, *35*, 1-248, 1993.

Fairbanks, R. G., A 17,000-year glacio-eustatic sealevel record: Influences of glacial melting rates on the Younger Dryas event and deep-ocean circulation, *Nature*, *342*, 637-642, 1989.

Fauquette, S., J. Guiot, M. Menut, J.-L. de Beaulieu, M. Reille, and P. Guenet, Vegetation and climate since the last interglacial in the Vienne area (France), *Global Planet. Change*, *20*, 1-17, 1999.

Fawcett, P. J., A. M. Ágústsdóttir, R. B. Alley, and C. A. Shuman, The Younger Dryas termination and North Atlantic Deep Water formation: Insights from climate model simulations and Greenland ice cores, *Paleoceanography*, *12*, 23-38, 1997.

Ferraz-Vicentini, K. R., and M. L. Salgado-Labouriau, Palynological analysis of a palm swamp in Central Brazil, *J. South Amer. Sci.*, *9*, 207-219, 1996.

Ferry, J. G., *Methanogenesis: Ecology, Physiology, Biochemistry and Genetics*, 536 pp., Chapman and Hall, New York, 1993.

Field, M. E., and K. A. Kvenvolden, Gas hydrates on the Northern California continental margin, *Geology*, *13*, 517-520, 1985.

Field, M. E., and J. H. Barber, A submarine landslide associated with shallow seafloor gas and gas hydrates off Northern California, in *Submarine Landslides: Selective Studies in the U. S. Exclusive Economic Zone*, edited by W. C. Schwab, H. J. Lee and D. C. Twichell, *U.S. Geological Survey Bulletin*, B2002, pp. 151-157, 1993.

Fisher, C. R., I. R. MacDonald, R. Sassen, C. M. Young, S. A. Macko, S. Hourdez, R. S. Carney, S. Joye, and E. McMullin, Methane ice worms: *Hesiocaeca methanicola* colonizing fossil fuel reserves, *Naturwissenschaften*, *87*, 184-187, 2000.

Fisk, H. N., and E. McFarlan, Jr., Late Quaternary deltaic deposits of the Mississippi River, *Geol. Soc. Am. Spec. Paper*, *62*, pp. 279-302, 1955.

Flückiger, J., A. Dällenbach, T. Blunier, B. Stauffer, T. F. Stocker, D. Raynaud, and J.-M. Barnola, Variations in atmospheric N_2O concentration during abrupt climatic changes, *Science*, *285*, 227-230, 1999.

Francois, R., M. P. Bacon, and D. O. Suman, Thorium-230 profiling in deep-sea sediments: High-resolution records of flux and dissolution of carbonate in the equatorial Atlantic during the last 24,000 years, *Paleoceanography*, *5*, 761-787, 1990.

Freeland, H. J., A. S. Bychkov, F. Whitney, C. Taylor, C. S. Wong, and G. I. Yurasov, WOCE section P1W in the Sea of Okhotsk - 1. Oceanographic data description, *J. Geophys. Res.*, *103*, 15613-15623, 1998.

Fung, I., J. John, J. Lerner, E. Matthews, M. Prather, L. P. Steele, and P. J. Fraser, Three-dimensional model synthesis of the global methane cycle, *J. Geophys. Res.*, *96*, 13033-13065, 1991.

Gajewski, K., A. Viau, M. Sawada, D. Atkinson, and S. Wilson, *Sphagnum* peatland distribution in North America and Eurasia during the past 21,000 years, *Global Biogeochem. Cycles*, *15*, 297-310, 2001.

Galchenko, V. F., A. Lein, and M. Ivanov, Biological sinks of methane, in *Exchange of Trace*

Gases Between Terrestrial Ecosystems and the Atmosphere, edited by M. O. Andreae and D. S. Schimel, pp. 39-58, John Wiley and Sons, New York, 1989.

Ganopolski, A., S. Rahmstorf, V. Petoukhov, and M. Claussen, Simulation of modern and glacial climates with a coupled global model of intermediate complexity, *Nature*, *391*, 351-356, 1998.

Gardner, J. V., Submarine geology of the western Coral Sea, *Geol. Soc. Am. Bull.*, *81*, 2599-2614, 1970.

Gardner, J. V., D. B. Prior, and M. E. Field, Humboldt slide—A large shear-dominated retrogressive slope failure, *Mar. Geol.*, *154*, 323-338, 1999.

Gardner, J. M., P. R. Vogt, and L. Somoza, The possible affect of the Mediterranean Outflow Water (MOW) on gas hydrate dissociation in the Gulf of Cadiz, *EOS Trans. AGU*, *82*(47), Fall Meet. Suppl., Abstract OS12B-0418, 2001.

Goodbred, S. L., Jr., and S. A. Kuehl, Enormous Ganges-Brahmaputra sediment discharge during strengthened early Holocene monsoon, *Geology*, *28*, 1083-1086, 2000.

Gore, A. J. P., and D. W. Goodall, (Eds.) *Mires: Swamp, Bog, Fen and Moor: General Studies, Ecosyst. World*, *4A*, 440 pp., Elsevier, New York, 1983a.

Gore, A. J. P., and D. W. Goodall, (Eds.) *Mires: Swamp, Bog, Fen and Moor: Regional Studies, Ecosyst. World*, *4B*, 479 pp., Elsevier, New York, 1983b.

Gorham, E., Northern peatlands: Role in the carbon cycle and probable responses to climatic warming, *Ecolog. Appl.*, *1*, 182-195, 1991.

Gorman, A. R., W. S. Holbrook, M. J. Hornbach, K. L. Hackwith, D. Lizarralde, and I. Pecher, Migration of methane gas through the hydrate stability zone in a low-flux hydrate province, *Geology*, *30*, 327-330, 2002.

Gornitz, V., and I. Fung, Potential distribution of methane hydrates in the world's oceans, *Global Biogeochem. Cycles*, *8*, 335-347, 1994.

Gould, S. J., *Wonderful Life: The Burgess Shale and the Nature of History*, 323 pp., W. W. Norton & Company, New York, 1989.

Greene, H. G., N. Maher, and C. K. Paull, Landslide hazards off of Santa Barbara, California, *EOS Trans. AGU*, *81*(48), Fall Meet. Suppl., OS21H-10, 2000.

Grigg, L., C. Whitlock, and W. E. Dean, Evidence for millennial-scale climate change during marine isotope stages 2 and 3 at Little Lake, western Oregon, U.S.A., *Quat. Res.*, *56*, 10-22, 2001.

Grimm, E. C., G. L. Jacobson, Jr., W. A. Watts, B. C. S. Hansen, and K. A. Maasch, A 50,000-year record of climate oscillations from Florida and its temporal correlation with Heinrich events, *Science*, *261*, 198-200, 1993.

Grootes, P. M., and M. Stuiver, Oxygen 18/16 variability in Greenland snow and ice with 10^{-3}- to 10^5-year time resolution, *J. Geophys. Res.*, *102*, 26455-26470, 1997.

Grootes, P. M., M. Stuiver, J. W. C. White, S. Johnsen, and J. Jouzel, Comparison of oxygen isotope records from the GISP2 and GRIP Greenland cores, *Nature*, *366*, 552-554, 1993.

Grootes, P. M., E. D. Waddington, E. J. Steig, D. L. Morse, M. Stuiver, and M. Nadeau, Global synchroneity of glacial climate fluctuations, *EOS Trans. AGU*, *81*(48), Fall Meet. Suppl., Abstract OS11E-02, 2000.

Grousett, F. E., L. Labeyrie, J. A. Sinko, M. Cremer, G. Bond, J. Duprat, E. Cortijo, and S. Huon, Patterns of ice-rafted detritus in the glacial North Atlantic, *Paleoceanography*, *8*, 175-192, 1993.

Guinasso, N. L., and D. R. Schink, A simple physicochemical acoustic model of methane bubbles rising in the sea: Report Reference 73-15-T, College of Geosciences, Texas A&M University, 1973.

Halsey, L. A., D. H. Vitt, and I. E. Bauer, Peatland initiation during the Holocene in continental western Canada, *Clim. Change*, *40*, 315-342, 1998.

Hampton, M. A., H. J. Lee, and J. Locat, Submarine landslides, *Rev. Geophys.*, *34*, 33-59, 1996.

Hanebuth, T. J. J., *Sea-Level Changes on the Sunda Shelf During the Last 50,000 Years*, 104 pp., Institut für Geowissenschaften, Christian-Albrechts-Universität, Kiel, 2000.

Hanebuth, T., and K. Stattegger, The stratigraphic evolution of the Sunda Shelf during the past fifty thousand years. SEPM Spec. Pub., in press.

Hanebuth, T., K. Stattegger, and P. M. Grootes, Rapid flooding of the Sunda Shelf: Late-glacial sea-level record, *Science*, *288*, 1033-1035, 2000.

Hansen, J., M. Sato, R. Ruedy, A. Lacis, and V. Oinas, Global warming in the twenty-first century: An alternative scenario, *Proc. Natl. Acad. Sci.*, *97*, 9875-9880, 2000.

Hanson, R. S., and T. E. Hanson, Methanotrophic bacteria, *Microbiol. Rev.*, *60*, 439-471, 1996.

Haq, B. U., Deep-sea response to eustatic change and significance of gas hydrates for continental margin stratigraphy, in *Sequence Stratigraphy and Facies Associations*, edited by H. W. Posamentier, C. P. Summerhayes, B. U. Haq and G. P. Allen, *Spec. Publ. Int. Ass. Sediment.*, *18*, pp. 93-106, 1993.

Haq, B. U., Growth and decay of gas hydrates: A forcing mechanism for abrupt climate change and sediment wasting on ocean margins?, *Koninklijke Akademie van Wetenschappen, Verhandelingen, Afd. Natuurkunde, Eerste Reeks, part 44*, 191-203, 1995.

Haq, B. U., Natural gas hydrates: Searching for the long-term climatic and slope-stability records, in *Gas Hydrates: Relevance to World Margin Stability and Climate Change*, edited by J.-P. Henriet and J. Mienert, *Geol. Soc. Spec. Publ.*, *137*, pp. 303-318, 1998a.

Haq, B. U., Gas hydrates: Greenhouse nightmare? Energy panacea or pipe dream?, *GSA Today*, *8*, 1-6, 1998b.

Haq, B. U., Methane in deep blue sea, *Science*, *285*, 543-544, 1999.

Haq, B. U., Climate impact of natural gas hydrates, in *Natural Gas Hydrate in Oceanic and Permafrost Environments*, edited by M. D. Max, pp. 137-148, Kluwer Academic Publishers, Dordrecht, Netherlands, 2000.

Harden, J. W., E. T. Sundquist, R. F. Stallard, and R. K. Mark, Dynamics of soil carbon during deglaciation of the Laurentide Ice Sheet, *Science*, *258*, 1921-1924, 1992.

Harris, P. T., C. B. Pattiaratchi, J. B. Keene, R. W. Dalrymple, J. V. Gardner, E. K. Baker, A. R. Cole, D. Mitchell, P. Gibbs, and W. W. Schroeder, Late Quaternary deltaic and carbonate sedimentation in the Gulf of Papua Foreland Basin: Response to sea-level change, *J. Sediment. Res.*, *66*, 801-819, 1996.

Harris, S. E., and A. C. Mix, Pleistocene precipitation balance in the Amazon Basin recorded in deep sea sediments, *Quat. Res.*, *51*, 14-26, 1999.

Harriss, R. C., E. Gorham, D. I. Sebacher, K. B. Bartlett, and P. A. Flebbe, Methane flux from northern peatlands, *Nature*, *315*, 652-654, 1985.

Harriss, R. C., D. I. Sebacher, K. B. Bartlett, D. S. Bartlett, and P. M. Crill, Sources of atmospheric methane in the south Florida environment, *Global Biogeochem. Cycles*, *2*, 231-243, 1988.

Harvey, L. D. D., and Z. Huang, Evaluation of the potential impact of methane clathrate destabilization on future global warming, *J. Geophys. Res.*, *100*, 2905-2926, 1995.

Hasiotis, T., G. Papatheodorou, N. Kastanos, and G. Ferentinos, A pockmark field in the Patras Gulf (Greece) and its activation during the 14/7/93 seismic event, *Mar. Geol.*, *130*, 333-344, 1996.

Hatzikiriakos, S. G., and P. Englezos, Permafrost melting and stability of offshore methane hydrates subject to global warming, *Int. J. Offshore Polar Eng.*, *4*, 162-166, 1994.

Haug, G. H., K. A. Hughen, D. M. Sigman, L. C. Peterson, and U. Röhl, Southward migration of the intertropical convergence zone through the Holocene, *Science*, *293*, 1304-1308, 2001.

Hay, W. W., and M. A. Leslie, Could possible changes in global groundwater reservoir cause eustatic sea-level fluctuations?, in *Sea-Level Change*, edited by R. Revelle, pp. 161-170, National Academy Press, Washington D.C., 1990.

Heinrich, H., Origin and consequences of cyclic ice rafting in the northeast Atlantic Ocean during the past 130,000 years, *Quat. Res.*, *29*, 143-152, 1988.

Hemming, S. R., P. E. Biscaye, W. S. Broecker, N. G. Hemming, M. Klas, and I. Hajdas, Provenance change coupled with increased clay flux during deglacial times in the western equatorial Atlantic, *Palaeogeogr. Palaeoclimatol. Palaeoecol.*, *142*, 217-230, 1998.

Hendy, I. L., and J. P. Kennett, Latest Quaternary north pacific surface-water responses imply atmospheric-driven climate instability, *Geology*, 27, 291-294, 1999.

Hendy, I. L., and J. P. Kennett, Dansgaard-Oeschger cycles and the California current system: Planktonic foraminiferal response to rapid climate change in Santa Barbara Basin, ocean drilling program hole 893A, *Paleoceanography*, *15*, 30-42, 2000.

Hendy, I. L., and J. P. Kennett, Tropical forcing of North Pacific Intermediate Water distribution during Late Quaternary rapid climate change?, *Quat. Sci. Rev.*, submitted.

Hendy, I. L., J. P. Kennett, E. B. Roark, and B. L. Ingram, Apparent synchroneity of submillennial scale climate events between Greenland and Santa Barbara Basin, California from 30-10 ka, *Quat. Sci. Rev.*, *21*, 1167-1184, 2002.

Henriet, J.-P., and J. Mienert, *Gas Hydrates: Relevance to World Margin Stability and Climatic Change: Tutorial Book: Het Pand*, 167 pp., Gent, Belgium, 1996.

Henriet, J.-P., and J. Mienert, (Eds.) *Gas Hydrates: Relevance to World Margin Stability and Climate Change, Geol. Soc. Spec. Publ*, *137*, 338 pp., 1998.

Hesse, R., and S. Khodabakhsh, Depositional facies of late Pleistocene Heinrich events in the Labrador Sea, *Geology*, *26*, 103-106, 1998.

Hesselbo, S. P., D. R. Grocke, H. C. Jenkyns, C. J. Bjerrum, P. Farrimond, H. S. Morgans Bell, and O. R. Green, Massive dissociation of gas hydrate during a Jurassic oceanic anoxic event, *Nature*, *406*, 392-395, 2000.

Hill, P. R., P. J. Mudie, K. Moran, and S. M. Blasco, A sea-level curve for the Canadian Beaufort Shelf, *Can. J. Earth Sci.*, *22*, 1383-1393, 1985.

Hill, P. R., A. Héquette, and M.-H. Ruz, Holocene sea-level history of Canadian Beaufort shelf, *Can. J. Earth Sci.*, *30*, 103-108, 1993.

Hill, T. M., and J. P. Kennett, Methane-associated foraminifera in Santa Barbara Basin, California: Ecology and stable isotopic composition, *EOS Trans. AGU*, *81*(48), Fall Meet. Suppl., Abstract T51B-10, 2000.

Hill, T. M., and J. P. Kennett, High resolution records of late Quaternary carbon isotopic excursions: Santa Barbara Basin, California, *EOS Trans. AGU*, *82*(47), Fall Meet. Suppl., Abstract PP12A-0483, 2001.

Hinnov, L. A., New perspectives on orbitally forced stratigraphy, *Annu. Rev. Earth Planet.*

Sci., *28*, 419-475, 2000.

Hinnov, L. A., M. Schulz, P. Yiou, Interhemispheric space-time attributes of the Dansgaard-Oeschger oscillations between 100 and 0 ka, *Quat. Sci. Rev.*, *21*, 1213-1228, 2002.

Hinrichs, K.-U., A molecular recorder of methane hydrate destabilization, *Geochem. Geophys. Geosyst.*, *2*, Paper number 2000GC000118, Jan. 9, 2001.

Hinrichs, K.-U., and A. Boetius, The anaerobic oxidation of methane: New insights in microbial ecology and biochemistry, in *Ocean Margin Systems*, edited by G. Wefer, D. Billet, D. Hebbeln, B. B. Jørgensen, M. Schlueter and T. V. Weering, Springer-Verlag, Heidelberg, in press.

Hinrichs, K.-U., J. M. Hayes, S. P. Sylva, P. G. Brewer, and E. F. DeLong, Methane-consuming archaebacteria in marine sediments, *Nature*, *398*, 802-805, 1999.

Hmelo, L., and K.-U. Hinrichs, Melting of marine hydrates in geologic history: What do molecular biomarkers tell us?, paper presented at Ocean Sciences Meeting, Honolulu, Hawaii, 11-15 February, 2002.

Hoehler, T. M., M. J. Alperin, D. B. Albert, and C. S. Martens, Field and laboratory studies of methane oxidation in anoxic marine sediments: Evidence for a methanogen-sulfate reducer consortium, *Global Biogeochem. Cycles*, *8*, 451-463, 1994.

Hoehler, T. M., M. J. Alperin, D. B. Albert, and C. S. Martens, Thermodynamic control on hydrogen concentrations in anoxic sediments, *Geochim. Cosomochim. Acta*, *62*, 1745-1756, 1998.

Hoelzmann, P., H.-J. Kruse, and F. Rottinger, Precipitation estimates for the eastern Saharan palaeomonsoon based on a water balance model of the West Nubian Palaeolake Basin, *Global Planet. Change*, *26*, 105-120, 2000.

Hoffman, P. F., G. P. Halverson, and J. P. Grotzinger, Are Proterozoic cap carbonates and isotopic excursions a record of gas hydrate destabilization following Earth's coldest intervals?: Comment, *Geology*, *30*, 286-287, 2002.

Holbrook, W. S., Seismic studies of the Blake Ridge: Implications for hydrate distribution, methane expulsion, and free gas dynamics, in *Natural Gas Hydrates: Occurrence, Distribution, and Detection*, edited by C. K. Paull and W. P. Dillon, *Geophys. Monogr. Ser.*, *124*, pp. 235-256, AGU, Washington D.C., 2001.

Holbrook, W. S., H. Hoskins, W. T. Wood, R. A. Stephen, D. Lizarralde, and Leg 164 Science Party, Methane hydrate and free gas on the Blake Ridge from vertical seismic profiling, *Science*, *273*, 1840-1843, 1996.

Hostetler, S. W., P. U. Clark, P. J. Bartlein, A. C. Mix, and N. J. Pisias, Atmospheric transmission of North Atlantic Heinrich events, *J. Geophys. Res.*, *104*, 3947-3952, 1999.

Hovland, M., and A. G. Judd, *Seabed Pockmarks and Seepages: Impact on Geology, Biology, and Marine Environment*, 293 pp., Graham and Trotman, Boston, 1988.

Hovland, M., A. G. Judd, and R. A. Burke, Jr., The Global flux of methane from shallow submarine sediments, *Chemosphere*, *26*, 559-578, 1993.

Hughen, K. A., J. T. Overpeck, J. T. Peterson, L. C. Peterson, and S. Trumbore, Rapid climate changes in the tropical Atlantic region during the last deglaciation, *Nature*, *380*, 51-54, 1996.

Hughen, K. A., J. T. Overpeck, S. J. Lehman, M. Kashgarian, J. Southon, L. C. Peterson, R. Alley, and D. M. Sigman, Deglacial changes in ocean circulation from an extended radiocarbon calibration, *Nature*, *391*, 65-68, 1998.

Hughen, K. A., J. R. Southon, S. J. Lehman, and J. T. Overpeck, Synchronous radiocarbon and climate shifts during the last deglaciation, *Science*, *290*, 1951-1954, 2000.

Hyndman, R. D., G. D. Spence, R. Chapman, M. Riedel, and R. N. Edwards, Geophysical studies of marine gas hydrate in northern Cascadia, in *Natural Gas Hydrates: Occurrence, Distribution, and Detection*, edited by C. K. Paull and W. P. Dillon, *Geophys. Monogr. Ser., 124*, pp. 273-296, AGU, Washington D.C., 2001.

Imbrie, J., E. A. Boyle, S. C. Clemens, A. Duffy, W. R. Howard, G. Kukla, J. Kutzbach, D. G. Martinson, A. McIntyre, A. C. Mix, B. Molfino, J. J. Morely, L. C. Peterson, N. G. Pisias, W. L. Prell, M. E. Raymo, N. J. Shackleton, and J. R. Toggweiler, On the structure and origin of major glaciation cycles. 1. Linear response to Milankovitch forcing, *Paleoceanography, 7*, 701-738, 1992.

Imbrie, J., A. Berger, and N. J. Shackleton, Role of orbital forcing: A two-million-year perspective, in *Global Changes in the Perspective of the Past*, edited by J. A. Eddy and H. Oeschger, pp. 263-277, Wiley, New York, 1993.

Ingram, B. L., and J. P. Kennett, Radiocarbon chronology and planktonic-benthic foraminiferal ^{14}C age differences in Santa Barbara Basin sediments, hole 893A, in *Proc. Ocean Drill. Program Sci. Results, 146, Part 2*, edited by J. P. Kennett, J. G. Baldauf and et al., pp. 19-27, Ocean Drilling Program, College Station, Texas, 1995.

Jahren, A. H., N. C. Arens, G. Sarmiento, J. Guerrero, and R. Amundson, Terrestrial record of methane hydrate dissociation in the Early Cretaceous, *Geology, 29*, 159-162, 2001.

Jones, G. A., and L. D. Keigwin, Evidence from the Fram Strait (78°N) for early deglaciation, *Nature, 336*, 57-59, 1988.

Jones, W. J., Diversity and physiology of methanogens, in *Microbial Production and Consumption of Greenhouse Gases: Methane, Nitrogen Oxides, and Halomethanes*, edited by J. E. Rogers and W. B. Whitman, pp. 39-55, American Society for Microbiology, Washington, D.C., 1991.

Jouzel, J., N. I. Barkov, J. M. Barnola, M. Bender, J. Chappellaz, C. Genthon, V. M. Kotlyakov, V. Lipenkov, C. Lorius, J. R. Petit, D. Raynaud, G. Raisbeck, C. Ritz, T. Sowers, M. Stievenard, F. Yiou, and P. Yiou, Extending the Vostok ice-core record of paleoclimate to the penultimate glacial peroid, *Nature, 364*, 407-412, 1993.

Jouzel, J., V. Masson, O. Cattani, S. Falourd, M. Stievenard, B. Stenni, A. Longinelli, S. J. Johnsen, J. P. Steffenssen, J. R. Petit, J. Schwander, R. Souchez, and N. I. Barkov, A new 27 ky high resolution East Antarctic climate record, *Geophys. Res. Lett., 28*, 3199-3202, 2001.

Judd, A. G., Geological sources of methane, in *Atmospheric Methane: Its Role in the Global Environment*, edited by M. A. K. Khalil, pp. 280-303, Springer-Verlag, Berlin, 2000.

Junk, W. J., Ecology of swamps on the middle Amazon, in *Mires: Swamp, Bog, Fen, and Moor: Regional Studies*, edited by A. J. P. Gore and D. W. Goodall, *Ecosyst. World, 4B*, pp. 269-294, Elsevier, New York, 1983.

Kaiho, K., T. Arinobu, R. Ishiwatari, H. Morgans, H. Okada, N. Takeda, N. Tazaki, G. Zhou, Y. Kajiwara, R. Matsumoto, A. Hirai, N. Niitsuma, and H. Wada, Latest Paleocene benthic foraminiferal extinction and environmental changes at Tawanui, New Zealand, *Paleoceanography, 11*, 447-465, 1996.

Kalin, R. M., and J. L. Jirikowic, A plausible hydrological scenario for the Bølling-Allerød atmospheric methane increase, *Holocene, 6*, 111-118, 1996.

Kanfoush, S. L., D. A. Hodell, C. D. Charles, T. P. Guilderson, P. G. Mortyn, and U. S. Ninnemann, Millennial-scale instability of the Antarctic ice sheet during the last glaciation, *Science, 288*, 1815-1818, 2000.

Kastner, M., Gas hydrates in convergent margins: Formation, occurrence, geochemistry, and global significance, in *Natural Gas Hydrates: Occurrence, Distribution, and Detection*, edited by C. K. Paull and W. P. Dillon, *Geophys. Monogr. Ser.*, *124*, pp. 67-86, AGU, Washington D.C., 2001.

Katz, M. E., D. K. Pak, G. R. Dickens, and K. G. Miller, The source and fate of massive carbon input during the latest Paleocene Thermal Maximum, *Science*, *286*, 1531-1533, 1999.

Katz, M. E., B. S. Cramer, G. S. Mountain, S. Katz, and K. G. Miller, Uncorking the bottle: What triggered the Paleocene/Eocene thermal maximum methane release?, *Paleoceanography*, *16*, 549- 562, 2001.

Kayen, R. E., and H. J. Lee, Pleistocene slope instability of gas hydrate-laden sediment on the Beaufort Sea margin, *Marine Geotechnol.*, *10*, 125-141, 1991.

Kayen, R. E., and H. J. Lee, Slope stability in regions of sea-floor gas hydrate: Beaufort Sea continental slope, in *Submarine Landslides: Selective Studies in the U. S. Exclusive Economic Zone*, edited by W. C. Schwab, H. J. Lee and D. C. Twichell, *U.S. Geological Survey Bulletin*, *B2002*, pp. 97-103, 1993.

Keeling, C. D., and T. P. Whorf, Possible forcing of global temperature by the oceanic tides, *Proc. Natl. Acad. Sci.*, *94*, 8321-8328, 1997.

Keeling, C. D., and T. P. Whorf, The 1,800-year oceanic tidal cycle: A possible cause of rapid climate change, *Proc. Natl. Acad. Sci.*, *97*, 3814-3819, 2000.

Keeling, R. F., and B. B. Stephens, Antarctic sea ice and the control of Pleistocene climate instability, *Paleoceanography*, *16*, 112-131, 2001.

Keigwin, L. D., Glacial-age hydrography of the far northwest Pacific Ocean, *Paleoceanography*, *13*, 323-339, 1998.

Keigwin, L. D., Late Pleistocene-Holocene paleoceanography and ventilation of the Gulf of California, *J. Oceanogr.*, *58*, 421-432, 2002.

Keigwin, L. D., and G. A. Jones, Deglacial climatic oscillations in the Gulf of California, *Paleoceanography*, *5*, 1009-1023, 1990.

Keigwin, L. D., and E. A. Boyle, Surface and deep ocean variability in the northern Sargasso Sea during marine isotope stage 3, *Paleoceanography*, *14*, 164-170, 1999.

Keller, M., and W. A. Reiners, Soil-atmosphere exchange of nitrous oxide, nitric oxide, and methane under secondary succession of pasture to forest in the Atlantic lowlands of Costa Rica, *Global Biogeochem. Cycles*, *8*, 399-409, 1994.

Kennedy, M. J., N. Christie-Blick, and L. Sohl, Are Proterozoic cap carbonates and isotopic excursions a record of gas hydrate destabilization following Earth's coldest interval?, *Geology*, *29*, 443-446, 2001.

Kennedy, M. J., N. Christie-Blick, and L. Sohl, Are Proterozoic cap carbonates and isotopic excursions a record of gas hydrate destabilization following Earth's coldest interval?: Reply, *Geology*, *30*, 287-288, 2002.

Kennett, D. M., and P. E. Hargraves, Benthic Diatoms and Sulfide Fluctuations: Upper Basin of Pettaquamscutt River, Rhode Island, *Estuarine Coastal Shelf Sci.*, *21*, 577-586, 1985.

Kennett, J. P., *Marine Geology*, 813 pp., Prentice-Hall, Englewood Cliffs, New Jersey, 1982.

Kennett, J. P., and L. Ingram, A 20,000-year record of ocean circulation and climate change from the Santa Barbara Basin, *Nature*, *377*, 510-514, 1995a.

Kennett, J. P., and L. Ingram, Paleoclimatic evolution of Santa Barbara Basin during the last 20 K.Y.: Marine evidence from hole 893A, in *Proc. Ocean Drill. Program Sci. Results, 146, Part 2*, edited by J. P. Kennett, J. G. Baldauf, et al., pp. 309-325, Ocean Drilling Program, College Station, Texas, 1995b.

Kennett, J. P., and L. C. Peterson, Rapid climate change: Ocean responses to Earth system instability in the late Quaternary, *The Legacy of the Ocean Drilling Program, JOIDES Journal 28, 5-9, 2002.*

Kennett, J. P., and C. C. Sorlien, Hypothesis for a bottom-simulating reflection in Holocene Strata, submarine slides, and catastrophic methane release, Santa Barbara Basin, California, *AAPG Bull., 83,* 692, 1998.

Kennett, J. P., and L. D. Stott, Abrupt deep-sea warming, paleoceanographic changes and benthic extinctions at the end of the Paleocene, *Nature, 353,* 225-229, 1991.

Kennett, J. P., and L. D. Stott, Terminal Paleocene mass extinction in the deep sea: Association with global warming, in *Effects of Past Global Change on Life,* edited by J. P. Kennett, J. G. Balduaf and M. Lyle, pp. 94-107, National Academy Press, Washington D.C., 1995.

Kennett, J. P., and K. Venz, Late Quaternary climatically related planktonic foraminiferal assemblage changes: Hole 893A. Santa Barbara Basin, California, in *Proc. Ocean Drill. Program Sci. Results, 146, Part 2,* edited by J. P. Kennett, J. G. Baldauf, et al., *281-293,* pp., Ocean Drilling Program, College Station, Texas, 1995.

Kennett, J. P., I. L. Hendy, and R. J. Behl, Late Quaternary foraminiferal carbon isotopic record in Santa Barbara Basin: Implications for rapid climate change, *EOS Trans. AGU,* 77(46), Fall Meet. Suppl., 294, 1996.

Kennett, J. P., K. G. Cannariato, I. L. Hendy, and R. J. Behl, Carbon isotopic evidence for methane hydrate instability during Quaternary interstadials, *Science, 288,* 128-133, 2000a.

Kennett, J. P., K. G. Cannariato, I. L. Hendy, and R. J. Behl, Role of methane hydrates in late Quaternary climate change (abstract), *EOS Trans. AGU, 81*(48), Fall Meet. Suppl., 630, 2000b.

Kennett, J. P., E. B. Roark, K. G. Cannariato, B. L. Ingram, and R. Tada, Latest Quaternary paleoclimatic and radiocarbon chronology, Hole 1017E, southern California margin, in *Proc. Ocean Drill. Program Sci. Results, 167,* edited by M. Lyle, I. Koizumi, C. Richter, et al., pp. 249-254, Ocean Drilling Program, College Station, Texas, 2000c.

Kennett, J. P., I. L. Hendy, and K. G. Cannariato, Timing of paleoceanographic changes associated with deglaciation off the Southern California Margin, in prep.

Kettunen, A., V. Kaitala, A. Lehtinen, A. Lohila, J. Alm, J. Silvola, and P. J. Martikainen, Methane production and oxidation potentials in relation to water table fluctuations in two boreal mires, *Soil Biol. Biochem., 31,* 1741-1749, 1999.

Kiefer, T., M. Sarnthein, H. Erlenkeuser, P. M. Grootes, and A. P. Roberts, North Pacific response to millennial-scale changes in ocean circulation over the last 60 kyr, *Paleoceanography, 16,* 178-189, 2001.

Kienast, S., and J. L. McKay, Sea surface temperatures in the subarctic Northeast Pacific reflect millennial-scale climate oscillations during the last 16 kyrs, *Geophys. Res. Lett.,* 28, 1563-1566, 2001.

Kienast, M., S. Steinke, K. Stattegger, and S. E. Calvert, Synchronous tropical South China Sea SST change and Greenland warming during deglaciation, *Science, 291,* 2132-2134, 2001.

Kleinberg, R. L., and P. G. Brewer, Probing gas hydrate deposits, *Amer. Sci., 89,* 244-251, 2001.

Kneller, M., and D. Peteet, Late Quaternary climate in the ridge and valley of Virginia, USA: Changes in vegetation and depositional environments, *Quat. Sci. Rev., 12,* 613-628, 1993.

Koch, P. L., J. Zachos, and P. Gingerich, Correlation between isotope records in marine and continental carbon reservoirs near the Paleocene/Eocene boundary, *Nature*, *358*, 319-322, 1992.

Kopp, H., BSR occurrence along the Sunda margin: Evidence from seismic data, *Earth Planet. Sci. Lett.*, *197*, 225-235, 2002.

Korhola, A., Holocene climatic variations in southern Finland reconstructed from peat-initiation data, *Holocene*, *5*, 43-58, 1995.

Kosters, E. C., and J. R. Suter, Facies relationships and systems tracts in the late Holocene Mississippi Delta Plain, *J. Sediment. Petrol.*, *63*, 727-733, 1993.

Kotilainen, A. T., and N. J. Shackleton, Rapid climate variability in the North Pacific Ocean during the past 95,000 years, *Nature*, *377*, 323-326, 1995.

Kovanen, D. J., and D. J. Easterbrook, Paleodeviations of radiocarbon marine reservoir values for the northeast Pacific, *Geology*, *30*, 243-246, 2002.

Kremenetski, C. V., T. Böttger, F. W. Junge, and A. G. Tarasov, Late- and postglacial environment of the Buzuluk area, middle Vulga region, Russia, *Quat. Sci. Rev.*, *18*, 1185-1203, 1999.

Kröhling, D. M., and M. Iriondo, Upper Quaternary paleoclimates of the Mar Chiquita area, North Pampa, Argentina, *Quat. Int.*, *57/58*, 149-163, 1999.

Krull, E. S., and G. J. Retallack, $\delta^{13}C$ depth profiles from paleosols across the Permian-Triassic boundary: Evidence for methane release, *Geol. Soc. Am. Bull.*, *112*, 1459-1472, 2000.

Krull, E. S., G. J. Retallack, H. J. Campbell, and G. L. Lyon, $\delta^{13}C_{org}$ chemostratigraphy of the Permian-Triassic boundary in the Maitai Group, New Zealand: evidence for high-latitudinal methane release, *New Zealand J. Geol. Geophys.*, *43*, 21-32, 2000.

Kukowski, N., and I. A. Pecher, Gas hydrates in nature: Results from geophysical and geochemical studies, *Mar. Geol.*, *164*, 1, 2000.

Kumar, N., and R. W. Embley, Evolution and origin of Ceará Rise: An aseismic rise in the western equatorial Atlantic, *Geol. Soc. Am. Bull.*, *88*, 683-694, 1977.

Kutzbach, J. E., and F. A. Street-Perrott, Milankovitch forcing of fluctuations in the level of tropical lakes from 18 to 0 kyr BP, *Nature*, *317*, 130-134, 1985.

Kvenvolden, K. A., Methane hydrates and global climate, *Global Biogeochem. Cycles*, *2*, 221-229, 1988a.

Kvenvolden, K. A., Methane hydrates—A major reservoir of carbon in the shallow geosphere?, *Chem. Geol.*, *71*, 41-51, 1988b.

Kvenvolden, K. A., Gas hydrates—Geological perspective and global change, *Rev. Geophys.*, *31*, 173-187, 1993.

Kvenvolden, K. A., A review of the geochemistry of methane in natural gas hydrate, *Org. Geochem.*, *23*, 997-1008, 1995.

Kvenvolden, K. A., G. D. Ginsburg, and V. A. Soloviev, Worldwide distribution of subaquatic gas hydrates, *Geo Mar. Lett.*, *13*, 32-40, 1993.

Kvenvolden, K. A., and A. Grantz, Gas hydrates of the Arctic Ocean region, in *The Arctic Ocean Region*, edited by A. Grantz, L. Johnson and J. F. Sweeney, *Geol. North Am.*, *L*, pp. 517-526, Geological Society of America, Boulder, Colorado, 1990.

Kvenvolden, K. A., and M. Kastner, Gas hydrates of the Peruvian outer continental margin, in *Proceedings of the Ocean Drilling Program, Scientific Results, v. 112*, pp. 517-526, 1990.

Kvenvolden, K. A., and T. D. Lorenson, The global occurrence of natural gas hydrate, in

Natural Gas Hydrates: Occurrence, Distribution, and Detection, edited by C. K. Paull and W. P. Dillon, *Geophys. Monogr. Ser.*, *124*, pp. 3-18, AGU, Washington D.C., 2001.

Kvenvolden, K. A., and M. A. McMenamin, Hydrates of natural gas: A review of their geologic occurrence, *U.S. Geol. Survey Circ.*, *C825*, 11, 1980.

Kvenvolden, K. A., T. D. Lorenson, and M. D. Lilley, Methane in the Beaufort Sea on the continental shelf of Alaska (abstract), *EOS Trans. AGU*, *73*(43), Fall Meet. Suppl., 309, 1992.

Kvenvolden, K. A., T. D. Lorenson, and W. S. Reeburgh, Attention turns to naturally occurring methane seepage, *EOS Trans. AGU*, *82*, 457, 2001.

Labeyrie, L., Glacial climate instability, *Science*, *290*, 1905-1907, 2000.

Labeyrie, L. D., J.-C. Duplessy, and P. L. Blanc, Variations in the mode of formation and temperature of oceanic deep waters over the past 125,000 years, *Nature*, *327*, 477-482, 1987.

Lachenbruch, A., J. Sass, B. Marshall, and T. Moses, Permafrost, heat flow, and the geothermal regime at Prudhoe Bay, Alaska, *J. Geophys. Res.*, *87*, 9301-9316, 1982.

Lagerklint, I. M., and J. D. Wright, Late glacial warming prior to Heinrich event 1: The influence of ice rafting and large ice sheets on the timing of initial warming, *Geology*, *27*, 1099-1102, 1999.

Lanoil, B. D., R. Sassen, M. T. La Duc, S. T. Sweet, and K. H. Nealson, Bacteria and Archaea physically associated with Gulf of Mexico gas hydrates, *Appl. Environ. Microbiol.*, *67*, 5143-5153, 2001.

Large, P. J., *Methylotrophy and Methanogenesis*, 88 pp., American Society for Microbiology, Wasington D.C., 1983.

Lassen, S., A. Kuijpers, H. Kunzendorf, H. Lindgren, J. Heinemeier, E. Jansen, and K. L. Knudsen, Intermediate water signal leads surface water response during Northeast Atlantic deglaciation, *Global Planet. Change*, *32*, 111-125, 2002.

Latrubesse, E. M., and E. Franzinell, Late Quaternary alluvial sedimentation in the upper Rio Negro basin, Amazonia, Brazil: Palaeohydrological implications, in *Palaeohydrology and Environmental Change*, edited by G. Benito, V. R. Baker and K. J. Gregory, pp. 259-271, Wiley and Sons, New York, 1998.

Lea, D. W., D. K. Pak, and H. J. Spero, Climate impact of Late Quaternary equatorial Pacific sea surface temperature variations, *Science*, *289*, 1719-1724, 2000.

Ledru, M.-P., J. Bertaux, and A. Sifeddine, Absence of last glacial maximum records in lowland tropical forests, *Quat. Res.*, *49*, 233-237, 1998a.

Ledru, M.-P., M. L. Salgado-Labouriau, and M. L. Lorscheitter, Vegetation dynamics in southern and central Brazil during the last 10,000 yr B.P., *Rev. Palaeobot. Palynol.*, *99*, 131-142, 1998b.

Ledru, M.-P., P. Mourguiart, G. Ceccantini, B. Turcq, and A. Sifeddine, Tropical climates in the game of two hemispheres revealed by abrupt climatic change, *Geology*, *30*, 275-278, 2002.

Lee, K. E., and N. C. Slowey, Cool surface waters of the subtropical North Pacific Ocean during the last glacial, *Nature*, *397*, 512-514, 1999.

Leggett, J., The nature of the greenhouse threat, in *Global Warming: The Greenpeace Report*, edited by J. Leggett, pp. 30, 40-41, Oxford University Press, New York, 1990.

Leifer, I., and J. Clark, Modeling trace gases in hydrocarbon seep bubbles. Application to marine hydrocarbon seeps in the Santa Barbara Channel, *Russian J. Geol. Geophys.*, in press.

Leifer, I., J. F. Clark, and R. F. Chen, Modifications of the local environment by natural marine hydrocarbon seeps, *Geophys. Res. Lett.*, *27*, 3711-3714, 2000.

Lerche, I., and E. Bagirov, Guide to gas hydrate stability in various geological settings, *Marine Petrol. Geol.*, *15*, 427-437, 1998.

Lewis, K. B., J.-Y. Collot, and S. E. Lallemand, The dammed Hikurangi Trough: A channel-fed trench blocked by subducting seamounts and their wake avalanches (New Zealand-France GeodyNZ Project), *Basin Res.*, *10*, 441-468, 1998.

Leyden, B. W., Late Quaternary aridity and Holocene moisture fluctuations in the Lake Valencia Basin, Venezuela, *Ecology*, *66*, 1279-1295, 1985.

Lézine, A. M., Late Quaternary vegetation and climate of the Sahel, *Quat. Res.*, *32*, 317-334, 1989.

Lézine, A. M., and H. Hooghiemstra, Land-sea comparison during the last glacial-interglacial transition: Pollen records from West Tropical Africa, *Palaeogeogr. Palaeoclimatol. Palaeoecol.*, *79*, 313-331, 1990.

Lin, H.-L., L. C. Peterson, J. T. Overpeck, and S. E. Trumbore, Late Quaternary climate change from $\delta^{18}O$ records of multiple species of planktonic foraminifera: High-resolution records from the anoxic Cariaco Basin, Venezuela, *Paleoceanography*, *12*, 415-427, 1997.

Liu, H.-S., Frequency variations of the Earth's obliquity and the 100-kyr ice-age cycles, *Nature*, *358*, 396-399, 1992.

Liu, H.-S., Glacial-interglacial changes induced by pulse modulation of the incoming solar radiation, *J. Geophys. Res.*, *103*, 26147-26164, 1998.

Liu, K.-B., Holocene paleoecology of the boreal forest and Great Lakes—St. Lawrence Forrest in Northern Ontario, *Ecolog. Monogr.*, *60*, 179-212, 1990.

Liu, K.-B., S. Sun, and X. Jiang, Environmental change in the Yangtze River Delta since 12,000 years B.P., *Quat. Res.*, *38*, 32-45, 1992.

Loehle, C., Geologic methane as a source for post-glacial CO_2 increases: The hydrocarbon pump hypothesis, *Geophys. Res. Lett.*, *20*, 1415-1418, 1993.

Long, D., S. Lammers, and P. Linke, Possible hydrate mounds within large sea-floor craters in the Barents Sea, in *Gas Hydrates: Relevance to World Margin Stability and Climate Change*, edited by J.-P. Henriet and J. Mienert, *Geol. Soc. Spec. Publ.*, *137*, pp. 223-237, 1998.

Lonsdale, P., A transform continental margin rich in hydrocarbons in the Gulf of California, *AAPG Bull.*, *69*, 1160-1180, 1985.

Lorius, C., J. Jouzel, D. Raynaud, J. Hansen, and H. Le Treut, The ice-core record: Climate sensitivity and future greenhouse warming, *Nature*, *347*, 139-145, 1990.

Lowell, T. V., C. J. Heusser, B. G. Anderson, P. I. Moreno, A. Heuser, L. E. Heusser, C. Schlucter, D. R. Marchant, and G. H. Denton, Interhemispheric correlation of Late Pleistocene glacial events, *Science*, *269*, 1541-1549, 1995.

Lund, D. C., and A. C. Mix, Millennial-scale deep water oscillations: Reflections of the north Atlantic in the deep Pacific from 10 to 60 ka, *Paleoceanography*, *13*, 1-19, 1998.

Luo, Y., H. Chen, G. Wu, and X. Sun, Records of natural fire and climate history during the last three glacial-interglacial cycles around the South China Sea—Charcoal record from the ODP 1144, *Sci. China, Ser. D*, *44*, 897-904, 2001.

Luyendyk, B., J. Kennett, and J. Clark, Increase in methane input to the atmosphere from hydrocarbon seeps on the world's continental shelves during lowered sea level, paper presented at American Association of Petroleum Geologists Hedberg Conference, 'Near-surface hydrocarbon migration: Mechanisms and seepage rates', Vancouver, BC, Canada, April 7-10, 2002.

Lynch-Stieglitz, J., R. G. Fairbanks, and C. D. Charles, Glacial-interglacial history of Antarctic intermediate water: Relative strengths of Antarctic versus Indian Ocean sources, *Paleoceanography*, 9, 7-29, 1994.

Lynch-Stieglitz, J., W. B. Curry, and N. Slowey, Weaker gulf stream in the Florida Straits during the Last Glacial Maximum, *Nature*, 402, 644-648, 1999.

Maas, A., A note on the formation of peat deposits in Indonesia, in *Tropical Lowland Peatlands of Southeast Asia*, edited by E. Maltby, C. P. Immirzi and R. J. Safford, pp. 97-99, IUCN, Gland, Switzerland, 1996.

MacAyeal, D. R., Binge/purge oscillations of the Laurentide ice sheet as a cause of the North Atlantic's Heinrich events, *Paleoceanography*, 8, 775-784, 1993.

MacDonald, G. J., *The Long-Term Impacts of Increasing Atmospheric Carbon Dioxide Levels*, 252 pp., Ballinger, Cambridge, Massachusetts, 1982.

MacDonald, G. J., Role of methane clathrates in past and future climates, *Clim. Change*, 16, 247-281, 1990a.

MacDonald, G. J., The future of methane as an energy resource, *Ann. Rev. Energy*, 15, 53-83, 1990b.

MacDonald, G. J., Clathrates, in *Encyclopedia of Earth System Science*, edited by W. A. Nierenberg, 1, pp. 475-484, 1992.

MacDonald, I. R., N. L. Guinasso, R. Sassen, J. M. Brooks, L. Lee, and K. T. Scott, Gas hydrate that breaches the sea floor on the continental slope of the Gulf of Mexico, *Geology*, 22, 699-702, 1994.

MacDonald, I. R., I. Leifer, R. Sassen, P. Stine, R. Mitchell, and N. Guinasso, Transfer of hydrocarbons from natural seeps to the water column and atmosphere, *Geofluids*, 2, 95-107, 2002.

MacKay, M. E., R. D. Jarrard, G. K. Westbrook, and R. D. Hyndman, Origin of bottom-simulating reflectors: Geophysical evidence from the Cascadia accretionary prism, *Geology*, 22, 459-462, 1994.

Majorowicz, J. A., and K. G. Osadetz, Gas hydrate distribution and volume in Canada, *AAPG Bull.*, 85, 1211-1230, 2001.

Maldonado, A., and C. H. Nelson, The Ebro margin study, northwestern Mediterranean Sea—An introduction, *Mar. Geol.*, 95, 157-163, 1990.

Malone, M. J., G. Claypool, J. B. Martin, and G. R. Dickens, Variable methane fluxes in shallow marine systems over geologic time: The composition and origin of pore waters and authigenic carbonates on the New Jersey shelf, *Mar. Geol.*, in press.

Maltby, E., C. P. Immirzi, and R. J. Safford, (Eds.) *Tropical Lowland Peatlands of Southeast Asia*, 294 pp., IUCN, Gland, Switzerland, 1996.

Manabe, S., and R. J. Stouffer, Coupled ocean-atmosphere model response to freshwater input: Comparison to Younger Dryas event, *Paleoceanography*, 12, 321-336, 1997.

Marchesi, J. R., A. J. Weightman, B. A. Cragg, R. J. Parkes, and J. C. Fry, Methanogen and bacterial diversity and distribution in deep gas hydrate sediments from the Cascadia Margin as revealed by 16S rRNA molecular analysis, *FEMS Microbiol. Ecol.*, 34, 221-228, 2001.

Marchitto, T. M., Jr., W. B. Curry, and D. W. Oppo, Millennial-scale changes in North Atlantic circulation since the last glaciation, *Nature*, 393, 557-561, 1998.

Martens, C. S., and R. A. Berner, Interstitial water chemistry of Long Island Sound sediments, I, dissolved gases, *Limnol. Oceanogr.*, *22*, 10-25, 1977.

Martin, L., J. Bertaux, T. Corrège, M.-P. Ledru, P. Mourguiart, A. Sifeddine, F. Soubiès, D. Wirrmann, K. Suguio, and B. Turcq, Astronomical forcing of contrasting rainfall changes in tropical South America between 12,400 and 8800 cal yr B.P., *Quat. Res.*, *47*, 117-122, 1997.

Maslin, M. A., and S. J. Burns, Reconstruction of the Amazon Basin effective moisture availability over the past 14,000 years, *Science*, *29*, 2285-2287, 2000.

Maslin, M., S. Burns, H. Erlenkeuser, and C. Hohnemann, Stable isotope records from sites 932 and 933, in *Proc. Ocean Drill. Program Sci. Results, 155*, edited by R. D. Flood, D. J. W. Piper, A. Klaus, L. C. Peterson, et al., pp. 305-318, Ocean Drilling Program, College Station, Texas, 1997.

Maslin, M., N. Mikkelsen, C. Vilela, and B. Haq, Sea-level and gas-hydrate-controlled catastrophic sediment failures of the Amazon Fan, *Geology*, *26*, 1107-1110, 1998.

Masson, D. G., Late Quaternary turbidity current pathways to the Madeira Abyssal Plain and some constraints on turbidity current mechanisms, *Basin Res.*, *6*, 17-33, 1994.

Matsumoto, R., Causes of the $\delta^{13}C$ anomalies of carbonates and a new paradigm 'Gas-Hydrate Hypothesis', *J. Soc. Japan*, *101*, 902-924, 1995.

Matthews, E., Wetlands, in *Atmospheric Methane: Its Role in the Global Environment*, edited by M. A. K. Khalil, pp. 202-233, Springer-Verlag, Berlin, 2000.

Matthews, E., and I. Fung, Methane emission from natural wetlands: Global distribution, area, and environmental characteristics of sources, *Global Biogeochem. Cycles*, *1*, 61-86, 1987.

Max, M. D., (Ed.) *Natural Gas Hydrate in Oceanic and Permafrost Environments*, 414 pp., Kluwer Academic Publishers, Dordrecht, Netherlands, 2000.

Mayewski, P. A., L. D. Meeker, M. S. Twickler, S. Whitlow, O. Yang, W. B. Lyons, and M. Prentice, Major features and forcing of high-latitude northern hemisphere atmospheric circulation using a 110,000-year-long glaciochemical series, *J. Geophys. Res.*, *102*, 26345-26366, 1997.

Mayle, F. E., R. Burbridge, and T. J. Killeen, Millennial-scale dynamics of southern Amazonian rain forests, *Science*, *290*, 2291-2294, 2000.

McAdoo, B. G., L. F. Pratson, and D. L. Orange, Submarine landslide geomorphology, US continental slope, *Mar. Geol.*, *169*, 103-136, 2000.

McCorkle, D. C., L. D. Keigwin, B. H. Corliss, and S. R. Emerson, The influence of micro-habitats on the carbon isotopic composition of deep sea benthic foraminifera, *Paleoceanography*, *5*, 161-185, 1990.

McCorkle, D. C., D. T. Heggie, and H. H. Veeh, Glacial and Holocene stable isotope distributions in the southeastern Indian Ocean, *Paleoceanography*, *13*, 20-34, 1998.

McGeary, D. F. R., and J. E. Damuth, Postglacial iron-rich crusts in hemipelagic deep-sea sediment, *Geol. Soc. Am. Bull.*, *84*, 1201-1212, 1973.

McGregor, B. A., and R. H. Bennett, Mass movement of sediment on the continental slope and rise seaward of the Baltimore Canyon Trough, *Mar. Geol.*, *33*, 163-174, 1979.

McIver, R. D., Hydrates of natural gas—Important agent in geological processes, *Geol. Soc. Am. Abstr. Prog.*, *9*, 1989-1990, 1977.

McIver, R. D., Role of naturally occurring gas hydrates in sediment transport, *AAPG Bull.*, *66*, 789-792, 1982.

McManus, J. F., D. W. Oppo, and J. L. Cullen, A 0.5-million year record of millennial-scale climate variability in the North Atlantic, *Science*, *283*, 971-975, 1999.

Meese, D., R. Alley, T. Gow, P. M. Grootes, P. Mayewski, M. Ram, K. Taylor, E. Waddington, and G. Zielinski, Preliminary depth-age scale of the GISP2 ice core, CRREL Special Report 94-1, 1994.

Merewether, R., M. S. Olsson, and P. Lonsdale, Accoustically detected hydrocarbon plumes rising from 2-km depths in Guaymas Basin, Gulf of California, *J. Geophys. Res.*, *90*, 3075-3085, 1985.

Mienert, J., and P. Bryn, Gas hydrate drilling conducted on the European margin, *EOS Trans. AGU*, *78*, 567-571, 1997.

Mienert, J., and J. Posewang, Evidence of shallow- and deep-water gas hydrate destabilizations in North Atlantic polar continental margin sediments, *Geo Mar. Lett.*, *19*, 143-149, 1999.

Milkov, A. V., and R. Sassen, Thickness of the gas hydrate stability zone, Gulf of Mexico continental slope, *Marine Petrol. Geol.*, *17*, 981-991, 2000.

Mitsch, W. J., and J. G. Gosselink, *Wetlands*, 722 pp., Van Nostrand Reinhold, New York, 1993.

Mix, A. C., J. Le, and N. J. Shackleton, Benthic foraminiferal stable isotope stratigraphy of Site 846: 0-1.8 Ma., in *Proc. Ocean Drill. Program Sci. Results, 138*, edited by N. G. Pisias, L. A. Mayer, T. R. Janecek, A. Palmer-Julson and T. H. van Andel, pp. 839-854, Ocean Drilling Program, College Station, Texas, 1995.

Mix, A. C., D. C. Lund, N. G. Pisias, P. Bodén, L. Bornmalm, M. Lyle, and J. Pike, Rapid climate oscillations in the northeast Pacific during the last deglaciation reflect northern and southern hemisphere sources, in *Mechanisms of Global Climate Change at Millennial Time Scales*, edited by P. U. Clark, R. S. Webb and L. D. Keigwin, *Geophy. Monogr. Ser.*, *112*, pp. 127-148, AGU, Washington D.C., 1999.

Monnin, E., A. Indermühle, A. Dällenbach, J. Flückiger, B. Stauffer, T. F. Stocker, D. Raynaud, and J.-M. Barnola, Atmospheric CO_2 concentrations over the last glacial termination, *Science*, *291*, 112-114, 2001.

Moodley, L., G. J. Van der Zwaan, P. M. J. Herman, A. J. Kempers, and P. van Breugel, Differential response of benthic meiofauna to anoxia with special reference to Foraminifera (Protista: Sarcodina), *Mar. Ecol. Prog. Ser.*, *158*, 151-163, 1997.

Moodley, L., G. J. Van der Zwaan, G. M. W. Rutten, R. C. E. Boom, and A. J. Kempers, Subsurface activity of benthic foraminifera in relation to porewater oxygen content: Laboratory experiments, *Mar. Micropaleontol.*, *34*, 91-106, 1998.

Moore, T. R., and R. Knowles, The Influence of water table levels on methane and carbon dioxide emissions from the peatland soils, *Can. J. Soil Sci.*, *69*, 33-38, 1989.

Mora, G., and L. M. Pratt, Isotopic evidence for cooler and drier conditions in the tropical Andes during the last glacial stage, *Geology*, *29*, 519-522, 2001.

Moreno, P. I., G. L. Jacobson, Jr., T. V. Lowell, and G. H. Denton, Interhemispheric climate links revealed by a late-glacial cooling episode in southern Chile, *Nature*, *409*, 804-808, 2001.

Mountain, G. S., Cenozoic margin construction and destruction offshore New Jersey, in *Timing and Depositional History of Eustatic Sequences: Constraints on Seismic Stratigraphy*, edited by C. A. Ross and D. Haman, *Cushman Foundation Spec. Publ.*, *24*, pp. 57-83, 1987.

Mountain, G. S., and B. E. Tucholke, Mesozoic and Cenozoic geology of the U.S. Atlantic continental slope and rise, in *Geologic Evolution of the United States Atlantic Margin*, edited by W. C. Poag, pp. 292-341, Van Nostrand Reinhold, New York, 1985.

Muhs, D. R., and M. Zárate, Late Quaternary eolian records of the Americas and their pale-oclimatic significance, in *Interhemispheric Climate Linkages*, edited by V. Markgraf, pp. 183-216, Academic Press, San Diego, 2001.

Mulitza, S., C. Rühlemann, T. Bickert, W. Hale, J. Pätzold, and G. Wefer, Late Quaternary $\delta^{13}C$ gradients and carbonate acumulation in the western equatorial Atlantic, *Earth Planet. Sci. Lett.*, *155*, 237-249, 1998.

Muller, R. A., and G. J. MacDonald, Glacial cycles and orbital inclination, *Nature*, *377*, 107-108, 1995.

Muller, R. A., and G. J. MacDonald, Glacial cycles and astronomical forcing, *Science*, *277*, 215-218, 1997.

Musashi, M., Y. Isozaki, T. Koike, and R. Kreulen, Stable carbon isotope signature in mid-Panthalassa shallow-water carbonates across the Permo-Triassic boundary: Evidence for ^{13}C-depleted superocean, *Earth Planet. Sci. Lett.*, *191*, 9-20, 2001.

Naqvi, S. W. A., D. A. Jayakumar, P. V. Narvekar, H. Nalk, V. V. S. S. Sarma, W. D'Souza, S. Joseph, and M. D. George, Increased marine production of N_2O due to intensifying anoxia on the Indian continental shelf, *Nature*, *408*, 346-349, 2000.

National Research Council (U.S.) Committee on Abrupt Climate Change, *Abrupt Climate Change: Inevitable Surprises*, 244 pp., National Academy Press, Washington D.C., 2002.

Neff, U., S. J. Burns, A. Mangini, M. Mudelsee, D. Fleitmann, and A. Matter, Strong coherence between solar variability and the monsoon in Oman between 9 and 6 kyr ago, *Nature*, *411*, 290-293, 2001.

Nelson, C. H., Late Pleistocene-Holocene transgressive sedimentation in deltaic and non-deltaic areas of the northeastern Bering epicontinental shelf, in *The Northeastern Bering Shelf: New Perspectives of Epicontinental Shelf Processes and Depositional Products*, edited by C. H. Nelson and S.-D. Nio, *Geologie en Mijnbouw*, *61*, pp. 5-18, De Bussy Ellerman Harms, Amsterdam, 1982.

Nelson, C. H., and S.-D. Nio, (Eds.) *The Northeastern Bering Shelf: New Perspectives of Epicontinental Shelf Processes and Depositional Products*, Geologie en Mijnbouw, *61*, 120 pp., De Bussy Ellerman Harms, Amsterdam, 1982.

Nelson, H., A. Maldonado, J. H. Barbar, and B. Alonso, Modern sand-rich and mud-rich Siliciclastic aprons: Alternative base-of-slope turbidite systems to submarine fans, in *Seismic Facies and Sedimentary Processes of Submarine Fans and Turbidite Systems*, edited by P. Weimer and M. H. Link, pp. 171-190, Springer-Verlag, Berlin, 1991.

Neustadt, M. I., Holocene peatland development, in *Late Quaternary Environments of the Soviet Union*, edited by A. A. Velichko, pp. 201-206, University of Minnesota Press, Minneapolis, 1984.

Newnham, R. M., and D. J. Lowe, Fine-resolution pollen record of late-glacial climate reversal from New Zealand, *Geology*, *28*, 759-762, 2000.

Ninnemann, U. S., and C. D. Charles, Regional differences in Quaternary subantarctic nutrient cycling: Link to intermediate and deep water ventilation, *Paleoceanography*, *12*, 560-567, 1997.

Ninnemann, U. S., C. D. Charles, and D. A. Hodell, Origin of global millennial scale climate events: Constraints from the southern ocean deep sea sedimentary record, in *Mechanisms of Global Climate Change at Millennial Time Scales*, edited by P. U. Clark, R. S. Webb and L. D. Keigwin, *Geophys. Monogr. Ser.*, *112*, pp. 99-112, AGU, Washington, D.C., 1999.

Nisbet, E. G., Some northern sources of atmospheric methane: Production, history, and future implications, *Can. J. Earth Sci.*, *26*, 1603-1611, 1989.

Nisbet, E., Climate change and methane, *Nature*, *347*, 23, 1990.

Nisbet, E. G., Sources of atmospheric CH_4 in early postglacial time, *J. Geophys. Res.*, *97*, 12859-12867, 1992.

Nisbet, E. G., and D. J. W. Piper, Giant submarine landslides, *Nature*, *392*, 329-330, 1998.

Norris, R. D., and U. Röhl, Carbon cycling and chronology of climate warming during the Palaeocene/Eocene transition, *Nature*, *401*, 775-778, 1999.

Nürnberg, D., R. Müller, and R. Schneider, Paleo-sea surface temperature calculations in the equatorial east Atlantic from Mg/Ca ratios in planktic foraminifera: A comparison to sea surface temperature estimates from $U^{k'}_{37}$, oxygen isotopes, and foraminiferal transfer function, *Paleoceanography*, *15*, 124-134, 2000.

O'Brien, S. R., P. A. Mayewski, L. D. Meeker, D. A. Meese, M. S. Twickler, and S. I. Whitlow, Compexity of Holocene climate as reconstructed from a Greenland ice core, *Science*, *270*, 1962-1964, 1995.

Oeschger, H., J. Beer, U. Siegenthaler, B. Stauffer, W. Dansgaard, C. C. Langway, Late glacial climate history from ice cores, in *Climate Processes and Climate Sensitivity*, edited by J. E. Hansen and T. Takahashi, *Geophys. Monogr. Ser.*, *29*, pp. 299-306, AGU, Washington, D.C., 1984.

O'Leary, D. W., The timing and spatial relations of submarine canyon erosion and mass movement on the New England continental slope and rise, in *Geology of the United States' Seafloor: The View From GLORIA*, edited by J. V. Gardner, M. E. Field and D. C. Twichell, pp. 47-58, Cambridge University Press, Cambridge, United Kingdom, 1996.

Oechel, W. C., and G. L. Vourlitis, The effects of climate change on land-atmosphere feedbacks in Arctic tundra regions, *Trends Ecol. Evol.*, *9*, 324-329, 1994.

Oechel, W. C., S. J. Hastings, G. Vourlitis, M. Jenkins, G. Reichers, and N. Grulke, Recent change of Arctic tundra ecosystems from a net carbon dioxide sink to a source, *Nature*, *361*, 520-523, 1993.

Oltmans, S. J., and D. J. Hofmann, Increase in lower-stratospheric water vapour at a mid-latitude northern hemisphere site from 1981 to 1994, *Nature*, *374*, 146-149, 1995.

Orphan, V. J., C. H. House, K.-U. Hinrichs, K. D. McKeegan, and E. F. DeLong, Methane-consuming Archaea revealed by directly coupled isotopic and phylogenetic analysis, *Science*, *293*, 484-487, 2001.

Ortiz, J., A. Mix, S. Harris, and S. O'Connell, Diffuse spectral reflectance as a proxy for percent carbonate content in North Atlantic sediments, *Paleoceanography*, *14*, 171-186, 1999.

Pachur, H.-J., S. Kröpelin, P. Hoelzmann, M. Goschin, and N. Altmann, Late Quaternary fluvio-lacustrine environments of western Nubia, *Berl. Geowiss. Abh.*, *120*, 203-260, 1990.

Padden, M., H. Weissert, and M. de Rafelis, Evidence for Late Jurassic release of methane from gas hydrate, *Geology*, *29*, 223-226, 2001.

Paillard, D., Glacial cycles: Toward a new paradigm, *Rev. Geophys.*, *39*, 325-346, 2001.

Paillard, D., and L. Labeyrie, Role of the thermohaline circulation in the abrupt warming after Heinrich events, *Nature*, *372*, 162-164, 1994.

Pálfy, J., A. Demény, J. Haas, M. Hetényi, M. J. Orchard, and I. Veto, Carbon isotope anomaly and other geochemical changes at the Triassic-Jurassic boundary from a marine section in Hungary, *Geology*, *29*, 1047-1050, 2001.

Parkes, R. J., B. A. Cragg, and P. Wellsbury, Recent studies on bacterial populations and processes in subseafloor sediments: A review, *Hydrogeol. J.*, *8*, 11-28, 2000.

Paull, C. K., and W. P. Dillon, (Eds.) *Natural Gas Hydrates: Occurrence, Distribution, and Detection, Geophys. Monogr. Ser., 124*, 315 pp., AGU, Washington D.C., 2001.

Paull, C. K., W. Ussler, III, and W. P. Dillon, Is the extent of glaciation limited by marine gas-hydrates?, *Geophys. Res. Lett., 18*, 432-434, 1991.

Paull, C. K., W. Ussler, III, and W. S. Borowski, Sources of biogenic methane to form marine gas hydrates, *Ann. N. Y. Acad. Sci., 715*, 392-409, 1994.

Paull, C. K., W. Ussler, III, W. S. Borowski, and F. N. Spiess, Methane-rich plumes on the Carolina continental rise: Associations with gas hydrates, *Geology, 23*, 89-92, 1995.

Paull, C. K., W. J. Buelow, W. Ussler, III, and W. S. Borowski, Increased continental-margin slumping frequency during sea-level lowstands above gas hydrate-bearing sediments, *Geology, 24*, 143-146, 1996.

Paull, C., W. Ussler, III, N. Maher, H. G. Greene, G. Rehder, T. Lorenson, and H. Lee, Pockmarks off Big Sur, California, *Mar. Geol., 181*, 323-335, 2002.

Pecher, I. A., N. Kukowski, C. R. Ranero, and R. von Huene, Gas hydrates along the Peru and middle America trench systems, in *Natural Gas Hydrates: Occurrence, Distribution, and Detection*, edited by C. K. Paull and W. P. Dillon, *Geophys. Monogr. Ser., 124*, pp. 257-271, AGU, Washington D.C., 2001.

Perry, C. A., and K. J. Hsu, Geophysical, archaeological, and historical evidence support a solar-output model for climatic change, *Proc. Natl. Acad. Sci., 97*, 12433-12438, 2000.

Peterson, K. M., W. D. Billings, and D. N. Reynolds, Influence of water table and atmospheric CO_2 concentration on the balance of Arctic tundra, *Arctic Alpine Res., 16*, 331-335, 1984.

Peterson, L. C., G. H. Haug, K. A. Hughen, and U. Röhl, Rapid changes in the hydrologic cycle of the tropical Atlantic during the last glacial, *Science, 290*, 1947-1951, 2000.

Petit, J. R., J. Jouzel, D. Raynaud, N. I. Barkov, J.-M. Barnola, I. Basile, M. Bender, J. Chappellaz, M. Davis, G. Delaygue, M. Delmotte, V. M. Kotlyakov, M. Legrand, V. Y. Lipenkov, C. Lorius, L. Pépin, C. Ritz, E. Saltzman, and M. Stievenard, Climate and atmospheric history of the past 420,000 years from the Vostok ice core, Antarctica, *Nature, 399*, 429-436, 1999.

Petit-Maire, N., Paleoclimates in the Sahara of Mali. A multidisciplinary study, *Episodes, 9*, 7-16, 1986.

Petit-Maire, N., Natural variability of the Earth's environments: The last two climatic extremes (18000±2000 and 8000±1000 yrs BP), *Sci. Terre Planetes, 328*, 273-279, 1999.

Petit-Maire, N., and J. Riser, Holocene paleohydrography of the Niger, in *Palaeoecology of Africa and the Surrounding Islands*, edited by J. A. Coetzee and E. M. Van Zinderen Bakker, pp. 135-141, A. A. Balkema, Rotterdam, 1981.

Petit-Maire, N., D. Commelin, J. Fabre, and M. Fontugne, First evidence for Holocene rainfall in the Tanezrouft hyperdesert and its margins, *Palaeogeogr. Palaeoclimatol. Palaeoecol., 79*, 333-338, 1990.

Petit-Maire, N., M. Fontugne, and C. Rouland, Atmospheric methane ratio and environmental changes in the Sahara and Sahel during the last 130 kyrs, *Palaeogeogr. Palaeoclimatol. Palaeoecol., 86*, 197-206, 1991.

Pierrehumbert, R. T., Subtropical water vapor as a mediator of rapid global climate change, in *Mechanisms of Global Climate Change at Millennial Time Scales*, edited by P. U. Clark, R. S. Webb and L. D. Keigwin, *Geophys. Monogr. Ser., 112*, pp. 339-361, AGU, Washington D.C., 1999.

Piper, D. J. W., and K. I. Skene, Latest Pleistocene ice-rafting events on the Scotian Margin (eastern Canada) and their relationship to Heinrich events, *Paleoceanography*, *13*, 205-214, 1998.

Piper, J. W., and W. R. Normark, Sandy fans-from Amazon to Heuneme and beyond, *AAPG Bull.*, *85*, 1407-1438, 2001.

Polak, B., Character and occurrence of peat deposits in the Malaysian tropics, in *Modern Quaternary Research in Southeast Asia*, edited by G.-J. Bartstra and W. A. Casperie, pp. 71-81, Balkema, Rotterdam, 1975.

Popenoe, P., and W. P. Dillon, Characteristics of the continental slope and rise off North Carolina from GLORIA and seismic-reflection data: The interaction of downslope and contour current processes, in *Geology of the United States' Seafloor: The View From GLORIA*, edited by J. V. Gardner, M. E. Field and D. C. Twichell, pp. 59-77, Cambridge University Press, Cambridge, United Kingdom, 1996.

Popenoe, P., E. A. Schmuck, and W. P. Dillon, The Cape Fear landslide: Slope failure associated with salt diapirism and gas hydrate decomposition, in *Submarine Landslides: Selected Studies in the U.S. Exclusive Economic Zone*, edited by W. C. Schwab, H. J. Lee and D. C. Twichell, *U.S. Geological Survey Bulletin*, *B2002*, pp. 40-53, 1993.

Porter, S. C., Hawaiian glacial ages, *Quat. Res.*, *12*, 161-187, 1979.

Porter, S. C., and Z. An, Correlation between climate events in the North Atlantic and China during the last deglaciation, *Nature*, *375*, 305-308, 1995.

Prado, J. L., and M. T. Alberdi, The mammalian record and climatic change over the last 30,000 years in the Pampean Region, Argentina, *Quat. Int.*, *57/58*, 165-174, 1999.

Prather, M. J., Lifetimes and eigenstates in atmospheric chemistry, *Geophys. Res. Lett.*, *21*, 801-804, 1994.

Prather, M. J., Time scales in atmospheric chemistry: Theory, GWPs for CH_4 and CO, and runaway growth, *Geophys. Res. Lett.*, *23*, 2597-2600, 1996.

Prather, M., R. Derwent, D. Ehhalt, P. Fraser, E. Sanhueza, and X. Zhou, Other trace gases and atmospheric chemistry, in *Climate Change 1994: Radiative Forcing of Climate Change and an Evaluation of the IPCC IS92 Emission Scenarios*, edited by J. T. Houghton, L. G. Meira Filho, J. Bruce, H. Lee, C. B. A., E. Haites, N. Harris and K. Maskell, pp. 73-126, Cambridge University Press, Cambridge, United Kingdom, 1995.

Prather, M., D. Ehhalt, D. F., R. Derwent, E. Dlugokencky, E. Holland, I. Isaksen, J. Katima, V. Kirchhoff, P. Matson, P. Midgley, and M. Wang, Atmospheric chemistry and greenhouse gases, in *Climate Change 2001: The Scientific Basis. Contribution of Working Group I to the Third Assessment Report of the Intergovernmental Panel on Climate Change*, edited by J. T. Houghton, Y. Ding, D. J. Griggs, M. Noguer, P. J. van der Linden, X. Dai, K. Maskell and C. A. Johnson, pp. 239-287, Cambridge Univesity Press, New York, 2001.

Pratt, L. M., A. M. Carmo, V. Brüchert, S. M. Monk, and J. M. Hayes, Episodically strong recycling of methane recorded by $^{13}C/^{12}C$, N/C, and H/C ratios for kerogens from the last 160,000 years in the Santa Barbara Basin at Hole 893, in *Proc. Ocean Drill. Program Sci. Results, 146, Part 2*, edited by J. P. Kennett, J. G. Baldauf and et al., pp. 213-218, Ocean Drilling Program, College Station, Texas, 1995.

Prentice, I. C., and I. Y. Fung, The sensitivity of terrestrial carbon storage to climate change, *Nature*, *346*, 48-51, 1990.

Prentice, I. C., and M. Sarthein, Self-regulatory processes in the biosphere in the face of climate change, in *Global Changes in the Perspective of the Past*, edited by J. A. Eddy and H. Oeschger, pp. 29-38, Wiley, New York, 1993.

Prinn, R. G., R. F. Weiss, B. R. Miller, J. Huang, F. N. Alyea, D. M. Cunnold, P. J. Fraser, D. E. Hartley, and P. G. Simmonds, Atmospheric trends and lifetime of CH_3CCL_3 and global OH concentrations, *Science, 269,* 187-192, 1995.

Prins, M. A., and G. Postma, Effects of climate, sea level, and tectonics unraveled for last deglaciation turbidite records of the Arabian Sea, *Geology, 28,* 375-378, 2000.

Quay, P. D., S. L. King, J. Stutsman, D. O. Wilbur, L. P. Steele, I. Fung, R. H. Gammon, T. A. Brown, G. W. Farwell, P. M. Grootes, and F. H. Schmidt, Carbon isotopic composition of atmospheric CH_4: Fossil and biomass burning source strengths, *Global Biogeochem. Cycles, 5,* 25-47, 1991.

Quinby-Hunt, M. S., and P. Wilde, Climate forcing by marine methane clathrates during the lower Paleozoic?, *EOS Trans. AGU, 76*(46), Fall Meet. Suppl., 306, 1995.

Rahmstorf, S., and R. Alley, Stochastic resonance in glacial climate, *EOS Trans. AGU, 83,* 129 and 135, 2002.

Rasmussen, R. A., M. A. K. Khalil, and F. Moraes, Permafrost methane content: 1. Experimental data from sites in northern Alaska, *Chemosphere, 26,* 591-594, 1993.

Rathbun, A. E., L. A. Levin, Z. Held, and K. C. Lohmann, Benthic foraminifera associated with cold methane seeps on the northern California margin: Ecology and stable isotopic composition, *Mar. Micropaleontol., 38,* 247-266, 2000.

Raymo, M. E., The timing of major climate terminations, *Paleoceanography, 12,* 577-585, 1997.

Raymo, M. E., K. Ganley, S. Carter, D. W. Oppo, and J. McManus, Millennial-scale climate instability during the early Pleistocene epoch, *Nature, 392,* 699-702, 1998.

Raynaud, D., and U. Siegenthaler, Role of trace gases: The problem of lead and lag, in *Global Changes in the Perspective of the Past,* edited by J. A. Eddy and H. Oeschger, pp. 173-188, Wiley, New York, 1993.

Raynaud, D., J. Jouzel, J.-M. Barnola, J. Chappellaz, R. J. Delmas, and C. Lorius, The ice record of greenhouse gases, *Science, 259,* 926-934, 1993.

Raynaud, D., J.-M. Barnola, J. Chappellaz, T. Blunier, A. Indermühle, and B. Stauffer, The ice record of greenhouse gases: A view in the context of future changes, *Quat. Sci. Rev., 19,* 9-17, 2000.

Reeburgh, W. S., Methane consumption in Cariaco Trench waters and sediments, *Earth Planet. Sci. Lett., 28,* 337-344, 1976.

Reeburgh, W. S., 'Soft Spots' in the global methane budget, in *Microbial Growth on C-1 Compounds,* edited by M. E. Lidstrom and F. R. Tabita, pp. 335-342, Kluwer Academic Publishers, Boston, 1996.

Reeburgh, W. S., and S. C. Whalen, High latitude ecosystems as CH_4 sources, *Ecolog. Bull., 42,* 62-70, 1992.

Reeburgh, W. S., B. B. Ward, S. C. Whalen, K. A. Sandbeck, K. A. Kilpatrick, and L. J. Kerkhof, Black Sea methane geochemistry, *Deep Sea Res., 38,* 1189-1210, 1991.

Reeburgh, W. S., S. C. Whalen, and M. J. Alperin, The role of methylotrophy in the global methane budget, in *Microbial Growth on C-1 Compounds,* edited by J. C. Murrell and D. P. Kelly, pp. 1-14, Intercept Ltd., Andover, U.K., 1993.

Rehder, G., R. S. Keir, E. Suess, and M. Rhein, Methane in the northern Atlantic controlled by microbial oxidation and atmospheric history, *Geophys. Res. Lett., 26,* 587-590, 1999.

Rehder, G., R. W. Collier, K. Heeschen, P. M. Kosro, J. Barth, and E. Suess, Enhanced marine CH_4 emissions to the atmosphere off Oregon caused by upwelling, *Global Biogeochem. Cycles,* 10.1029/2000GBØØ1391, 2002.

Reichart, G. J., L. J. Lourens, and W. J. Zachariasse, Temporal variability in the northern Arabian Sea Oxygen Minimum Zone (OMZ) during the last 225,000 years, *Paleoceanography*, *13*, 607-621, 1998.

Reid, J. L., *Intermediate Waters of the Pacific Ocean*, 85 pp., John Hopkins Press, Baltimore, 1965.

Retallack, G. J., A 300-million-year record of atmospheric carbon dioxide from fossil plant cuticles, *Nature*, *411*, 287-290, 2001.

Revelle, R. R., Methane hydrates in continental slope sediments and increasing atmospheric carbon dioxide, in *Changing Climate: Report of the Carbon Dioxide Assessment Committee*, pp. 252-261, National Academy Press, Washington D.C., 1983.

Rial, J. A., Pacemaking the ice ages by frequency modulation of Earth's orbital eccentricity, *Science*, *285*, 564-568, 1999.

Richardson, D., The origin of iron-rich layers in sediments of the Western Equatorial Atlantic Ocean, Ph.D. thesis, Columbia University, New York, 1974.

Rieley, J. O., S. Page, and G. Sieffermann, Tropical peat swamp forests of southeast Asia: Ecology and environmental importance, *Malaysian J. Tropical Geog.*, *26*, 131-141, 1995.

Roberts, H. H., Fluid and gas expulsion on the northern Gulf of Mexico continental slope: Mud-prone to mineral-prone responses, in *Natural Gas Hydrates: Occurrence, Distribution, and Detection*, edited by C. K. Paull and W. P. Dillon, *Geophys. Monogr. Ser.*, *124*, pp. 145-161, AGU, Washington D.C., 2001.

Roberts, H. H., and P. Aharon, Hydrocarbon-derived carbonate buildups of the northern Gulf of Mexico continental slope: A review of submersible investigations, *Geo Mar. Lett.*, *14*, 135-148, 1994.

Roberts, H. H., and J. M. Coleman, Holocene evolution of the deltaic plain: A perspective—From Fisk to present, *Eng. Geol.*, *45*, 113-138, 1996.

Roberts, H. H., W. J. Wiseman, Jr., J. Hooper, and G. D. Humphrey, Surficial gas hydrates of the Louisiana continental slope: Initial results of direct observations and *in situ* data collection, in *31st Annual Offshore Technology Conference: Proceedings*, Proceedings—Offshore Technology Conference, *31*, pp. 259-272, 1999.

Rodhe, H., A comparison of the contribution of various gases to the greenhouse effect, *Science*, *248*, 1217-1219, 1990.

Rodriguez, N. M., and C. K. Paull, Data Report: [14]C dating of sediment of the uppermost Cape Fear Slide Plain: Constraints on the timing of this massive submarine landslide, in *Proc. Ocean Drill. Program Sci. Results, 164*, edited by C. K. Paull, R. Matsumoto, P. J. Wallace and W. P. Dillon, pp. 325-327, Ocean Drilling Program, College Station, Texas, 2000.

Rogers, J. E., and W. B. Whitman, (Eds.) *Microbial Production and Consumption of Greenhouse Gases: Methane, Nitrogen Oxides, and Halomethanes*, 298 pp., American Society for Microbiology, Washington, D.C., 1991.

Röhl, U., T. J. Bralower, R. D. Norris, and G. Wefer, A new chronology for the late Paleocene thermal maximum and its environmental implications, *Geology*, *28*, 927-930, 2000.

Romanowicz, E. A., D. I. Siegel, J. P. Chanton, and P. H. Glaser, Temporal variations in dissolved methane deep in the Lake Agassiz Peatlands, Minnesota, *Global Biogeochem. Cycles*, *9*, 197-212, 1995.

Rothwell, R. G., T. J. Pearce, and P. P. E. Weaver, Late Quaternary evolution of the Madeira Abyssal Plain, Canary Basin, NE Atlantic, *Basin Res.*, *4*, 103-131, 1992.

Rothwell, R. G., J. Thomson, and G. Kähler, Low-sea-level emplacement of a very large Late

Pleistocene 'megaturbidite' in the western Mediterranean Sea, *Nature*, *392*, 377-380, 1998.

Roulet, N. T., R. Ash, and T. R. Moore, Low boreal wetlands as a source of atmospheric methane, *J. Geophys. Res.*, *97*, 3739-3749, 1992.

Roulet, N. T., R. Ash, W. Quinton, and T. Moore, Methane flux from drained nothern peatlands: Effect of a persistent water table lowering on flux, *Global Biogeochem. Cycles*, *7*, 749-769, 1993.

Roulet, N. T., A. Jano, C. A. Kelly, L. F. Klinger, T. R. Moore, R. Protz, J. A. Ritter, and W. R. Rouse, Role of the Hudson Bay lowland as a source of atmospheric methane, *J. Geophys. Res.*, *99*, 1439-1454, 1994.

Ruddiman, W. F., Tropical Atlantic terrigenous fluxes since 25,000 yrs B.P., *Mar. Geol.*, *136*, 189-207, 1997.

Ruddiman, W. F., and F. A. Bowles, Early interglacial bottom-current sedimentation on the eastern Reykjanes Ridge, *Mar. Geol.*, *21*, 191-210, 1976.

Ruddiman, W. F., J. S. Thomson, The case for human causes of increased atmospheric CH_4, *Quat. Sci. Rev.*, *20*, 1769-1777, 2001

Rühlemann, C., S. Mulitza, P. J. Müller, G. Wefer, and R. Zahn, Warming of the tropical Atlantic Ocean and slowdown of thermohaline circulation during the last deglaciation, *Nature*, *402*, 511-514, 1999.

Rühlemann, C., B. Diekmann, S. Mulitza, and M. Frank, Late Quaternary changes of western equatorial Atlantic surface circulation and Amazon lowland climate recorded in Ceará Rise deep-sea sediments, *Paleoceanography*, *16*, 293-305, 2001.

Ruppel, C., Anomalously cold temperatures observed at the base of the gas hydrate stability zone on the U.S. Atlantic passive margin, *Geology*, *25*, 699-702, 1997.

Ruppel, C., and M. Kinoshita, Fluid, methane, and energy flux in an active margin gas hydrate province, offshore Costa Rica, *Earth Planet. Sci. Lett.*, *179*, 153-165, 2000.

Ruuhijarvi, R., The Finnish mire types and their regional distribution, in *Mires: Swamp, Bog, Fen, and Moor: Regional Studies*, edited by A. J. P. Gore and D. W. Goodall, *Ecosyst. World*, *4B*, pp. 47-67, Elsevier, New York, 1983.

Sachs, J. P., and S. J. Lehman, Subtropical North Atlantic temperatures 60,000 to 30,000 years ago, *Science*, *286*, 756-759, 1999.

Salgado-Labouriau, M. L., M. Barberi, K. R. Ferraz-Vicentini, and M. G. Parizzi, A dry climatic event during the late Quaternary of tropical Brazil, *Rev. Palaeobot. Palynol.*, *99*, 115-129, 1998.

Sansone, F. J., B. N. Popp, A. Gasc, A. W. Graham, and T. M. Rust, Highly elevated methane in the eastern tropical North Pacific and associated isotopically enriched fluxes to the atmosphere, *Geophys. Res. Lett.*, *28*, 4567-4570, 2001.

Sarnthein, M., and L. Diester-Haass, Eolian-Sand Turbidites, *J. Sediment. Petrol.*, *47*, 868-890, 1977.

Sarnthein, M., J. Thiede, U. Pflaumann, H. Erlenkeuser, D. Fütterer, B. Koopman, H. Lange, and E. Seibold, Atmospheric and oceanic circulation patterns off Northwest Africa during the past 25 million years, in *Geology of the Northwest African Continental Margin*, edited by U. von Rad, K. Hinz, M. Sarnthein and E. Seibold, pp. 545-604, Springer-Verlag, Berlin, 1982.

Sarnthein, M., J. P. Kennett, J. Chappell, T. Crowley, W. Curry, J. C. Duplessy, P. Grootes, I. Hendy, C. Laj, J. Negendank, M. Schulz, N. J. Shackleton, A. Voelker, B. Zolitschka, and the other Trins workshop participants, Exploring Late Pleistocene climate variations,

EOS Trans. AGU, *81*, 625 and 629-630, 2000.

Sarnthein, M., K. Stattegger, D. Dreger, H. Erlenkeuser, P. Grootes, B. Haupt, S. Jung, T. Kiefer, W. Kuhnt, U. Pflaumann, Schäfer-Neth, H. Schulz, M. Schulz, D. Seidov, J. Simstich, S. van Kreveld, E. Vogelsang, A. Völker, and M. Weinelt, Fundamental modes and abrupt changes in North Atlantic circulation and climate over the last 60 ky—concepts, reconstruction, and numerical modelling, in *The Northern North Atlantic: A Changing Environment*, edited by P. Schäfer, et al., pp., Springer-Verlag, Berlin, in press.

Sassen, R., S. T. Sweet, D. A. DeFreitas, A. V. Milkov, G. G. Salata, and E. W. McDade, Geology and geochemistry of gas hydrates, Central Gulf of Mexico continental slope, *AAPG Bull.*, *83*, 1363, 1999.

Sassen, R., S. T. Sweet, A. V. Milkov, D. A. DeFreitas, and M. C. Kennicutt, Thermogenic vent gas and gas hydrate in the Gulf of Mexico slope: Is gas hydrate decomposition significant?, *Geology*, *29*, 107-110, 2001a.

Sassen, R., S. T. Sweet, A. V. Milkov, D. A. DeFreitas, M. C. Kennicutt, and H. H. Roberts, Stability of thermogenic gas hydrate in the Gulf of Mexico: Constraints on models of climate change, in *Natural Gas Hydrates: Occurrence, Distribution, and Detection*, edited by C. K. Paull and W. P. Dillon, *Geophys. Monogr. Ser.*, *124*, pp. 273-296, AGU, Washington D.C., 2001b.

Schimel, J., Trace gases, soil, in *Encyclopedia of Environmental Microbiology*, edited by G. Britton, pp., John Wiley & Sons, New York, 3183-3184, 2001.

Schimel, J. P., E. A. Holland, and D. Valentine, Controls on methane flux from terrestrial ecosystems, in *Agroecosystem Effects on Radiatively Active Trace Gases and Global Climate Change*, edited by A. R. Mosier, J. Duxbury and L. Harper, *Am. Soc. Agron. Spec. Publ.*, *55*, pp. 167-182, American Society of Agronomy, Crop Science Society of America and Soil Science Society of America,, Madison, Wisconsin, 1993.

Schlesinger, W. H., Evidence from chronosequence studies for a low carbon storage potential of soils, *Nature*, *348*, 232-234, 1990.

Schmidt, G. A., and D. T. Shindell, Atmospheric composition, radiative forcing and climate change as a consequence of a catastrophic methane hydrate release, *EOS Trans. AGU*, *82*(47), Fall Meet. Suppl., Abstract PP32A-0510, 2001.

Schmitz, B., Plankton cooled a greenhouse, *Nature*, *407*, 143-174, 2000.

Schubert, C., Climatic changes during the Last Glacial Maximum in northern South America and the Caribbean: A review, *Interciencia*, *13*, 128-137, 1988.

Schubert, C. J., J. Villanueva, S. E. Calvert, G. L. Cowie, U. von Rad, H. Schulz, U. Berner, and H. Erlenkeuser, Stable phytoplankton community structure in the Arabian Sea over the past 200,000 years, *Nature*, *394*, 563-566, 1998.

Schulte, S., F. Rostek, E. Bard, J. Rullkötter, and O. Marchal, Variations of oxygen-minimum and primary productivity recorded in sediments of the Arabian Sea, *Earth Planet. Sci. Lett.*, *173*, 205-221, 1999.

Schulz, E., and J. W. Whitney, Upper Pleistocene and Holocene lakes in the An Nafud, Saudi Arabia, *Hydrobiologia*, *143*, 175-190, 1986.

Schulz, H., U. von Rad, and H. Erlenkeuser, Correlation between Arabian Sea and Greenland climate oscillations of the past 110,000 years, *Nature*, *393*, 54-57, 1998.

Schumm, S. A., River response to baselevel change: Implications for sequence stratigraphy, *J. Geol.*, *101*, 279-294, 1993.

Schütz, H., W. Seiler, and H. Rennenberg, Soil and land use related sources and sinks of methane (CH_4) in the context of the global methane budget, in *Soils and the Greenhouse*

Effect, edited by A. F. Bouwman, pp. 271-285, J. Wiley and Sons, New York, 1990.

Schwander, J., Gas diffusion in firn, in *Chemical Exchange Between the Atmosphere and Polar Snow*, edited by E. W. Wolff and R. C. Bales, *NATO ASI Ser.*, *143*, pp. 527-540, Springer-Verlag, New York, 1996.

Schwander, J., J.-M. Barnola, C. Andrié, M. Leuenberger, A. Ludin, D. Raynaud, and J. Stauffer, The age of the air in the firn and the ice at Summit, Greenland, *J. Geophys. Res.*, *98*, 2831-2838, 1993.

Scourse, J. D., J. Young, N. McCave, and C. Sugdon, Synchronous deposition of ice-rafted detritus from Laurentide and British ice sheet sources during Heinrich event 2 (abstract), *EOS Trans. AGU*, *80*, Fall Meet. Suppl., 1, 1999.

Scranton, M. I., and P. G. Brewer, Consumption of dissolved methane in the deep ocean, *Limnol. Oceanogr.*, *23*, 1207-1213, 1978.

Seidov, D., and M. Maslin, North Atlantic deep water circulation collapse during Heinrich events, *Geology*, *27*, 23-26, 1999.

Seltzer, G., D. Rodbell, and S. Burns, Isotopic evidence for late Quaternary climatic change in tropical South America, *Geology*, *28*, 35-38, 2000.

Severinghaus, J. P., and E. J. Brook, Abrupt climate change at the end of the last glacial period inferred from trapped air in polar ice, *Science*, *286*, 930-933, 1999.

Severinghaus, J. P., T. Sowers, E. J. Brook, R. B. Alley, and M. L. Bender, Timing of abrupt climate change at the end of the Younger Dryas interval from thermally fractionated gases in polar ice, *Nature*, *391*, 141-146, 1998.

Shackleton, N., Oxygen isotopes, ice volume and sea level, *Quat. Sci. Rev.*, *6*, 183-190, 1987.

Shackleton, N. J., The 100,000-year ice-age cycle identified and found to lag temperature, carbon dioxide, and orbital eccentricity, *Science*, *289*, 1897-1902, 2000.

Shackleton, N. J., Climate change across the hemispheres, *Science*, *291*, 58-59, 2001.

Shackleton, N. J., and M. A. Hall, Phase relationships between millennial-scale events 64,000-24,000 years ago, *Paleoceanography*, in press.

Shackleton, N. J., M. A. Hall, and D. Pate, Pliocene stable isotope stratigraphy of Site 846, in *Proc. Ocean Drill. Program Sci. Results, 138*, edited by N. G. Pisias, L. A. Mayer, T. R. Janecek, A. Palmer-Julson and T. H. van Andel, pp. 337-355, Ocean Drilling Program, College Station, Texas, 1995.

Shiga, K., and I. Koizumi, Latest Quaternary oceanographic changes in the Okhotsk Sea based on diatom records, *Mar. Micropaleontol.*, *38*, 91-117, 2000.

Short, D. A., and J. G. Mengel, Tropical climatic phase lags and Earth's precession cycle, *Nature*, *323*, 48-50, 1986.

Siani, G., M. Paterne, E. Michel, R. Sulpizio, A. Sbrana, M. Arnold, and G. Haddad, Mediterranean sea surface radiocarbon reservoir age changes since the last glacial maximum, *Science*, *294*, 1917-1920, 2001.

Siegert, M. J., J. A. Dowdeswell, J.-I. Svendsen, and A. Elverhøi, The Eurasian Arctic during the last ice age, *Amer. Sci.*, *90*, 32-39, 2002.

Sigman, D. M., and E. A. Boyle, Glacial/interglacial variations in atmospheric carbon dioxide, *Nature*, *407*, 859-869, 2000.

Sirocko, F., M. Sarnthein, H. Erlenkeuser, H. Lange, M. Arnold, and J.-C. Duplessy, Century-scale events in monsoonal climate over the past 24,000 years, *Nature*, *364*, 322-324, 1993.

Sirocko, F., D. G. Schönberg, and C. Devey, Processes controlling trace element geochemistry of Arabian Sea sediments during the last 25,000 years, *Global Planet. Change*, *26*, 217-303, 2000.

Sloan, E. D., Jr., *Clathrate Hydrates of Natural Gases*, 641 pp., Marcel Dekker, New York, 1990.

Sloan, E. D., *Clathrate Hydrates of Natural Gases*, 2 ed., 705 pp., Marcel Dekker, New York, 1998.

Slowey, N. C., and W. B. Curry, Glacial-interglacial differences in circulation and carbon cycling within the upper western North Atlantic, *Paleoceanography*, *10*, 715-732, 1995.

Smith, L. C., G. A. MacDonald, K. E. Frey, A. Velichko, K. Kremenetski, O. Borisova, P. Dubinin, and R. R. Forster, U.S.-Russia venture probes Siberian peatlands' sensitivity to climate, *EOS Trans. AGU*, *81*, 497 and 503-504, 2000.

Smith, L. M., J. P. Sachs, A. E. Jennings, D. M. Anderson, and A. deVernal, Light $\delta^{13}C$ events during deglaciation of the East Greenland continental shelf attributed to methane release from gas hydrates, *Geophys. Res. Lett.*, *28*, 2217-2220, 2001.

Solheim, A., and A. Elverhøi, Gas-related sea floor craters in the Barents Sea, *Geo Mar. Lett.*, *13*, 235-243, 1993.

Sowers, T., and M. Bender, Climate records covering the last deglaciation, *Science*, *269*, 210-214, 1995.

Sowers, T., M. Bender, L. Labeyrie, D. Martinson, J. Jouzel, D. Raynaud, J. J. Pichon, and Y. Korotkevich, 135,000 year Vostok-SPECMAP common temporal framework, *Paleoceanography*, *8*, 737-766, 1993.

Spence, G. D., N. R. Chapman, R. D. Hyndman, and C. Cleary, Fishing trawler nets massive "catch" of methane hydrates, *EOS Trans. AGU*, *82*, 621 and 627, 2001.

Stanley, D. J., Dating modern deltas: Progress, problems, and prognostics, *Annu. Rev. Earth Planet. Sci.*, *29*, 257-294, 2001.

Stanley, D. J., and A. G. Warne, Worldwide initiation of Holocene marine deltas by deceleration of sea-level rise, *Science*, *265*, 228-232, 1994.

Stanley, D. J., and A. K. Hait, Deltas, radiocarbon dating, and measurements of sediment storage and subsidence, *Geology*, *28*, 295-298, 2000.

Stattegger, K., W. Kuhnt, H. K. Wong, C. Bühring, C. Haft, T. Hanebuth, H. Kawamura, M. Kienast, S. Lorenc, B. Lotz, and T. Lüdmann, *Cruise Report SONNE 115 Sundaflut: Sequence Stratigraphy, Late Pleistocene-Holocene Sea Level Fluctuations and High Resolution Record of the Post-Pleistocene Transgression on the Sunda Shelf: Kota Kinabalu-Singapore, December 13, 1996-January 25, 1997*, 211 pp., Christian-Albrechts-Universität, Kiel, 1997.

Stauffer, B., T. Blunier, A. Dällenbach, A. Indermühle, J. Scwander, T. F. Stocker, J. Tschumi, J. Chappellaz, D. Raynaud, C. U. Hammer, and H. B. Claussen, Atmospheric CO_2 concentration and millennial-scale climate change during the last glacial period, *Nature*, *392*, 59-62, 1998.

Steig, E. J., E. J. Brook, J. W. C. White, C. M. Sucher, M. L. Bender, S. J. Lehman, D. L. Morse, E. D. Waddington, and G. D. Clow, Synchronous climate changes in Antarctica and the North Atlantic, *Science*, *282*, 92-95, 1998.

Stenni, B., V. Masson-Delmotte, S. Johnsen, J. Jouzel, A. Longinelli, E. Monnin, R. Röthlisberger, and E. Selmo, An oceanic cold reversal during the last deglaciation, *Science*, *293*, 2074-2077, 2001.

Stevaux, J. C., The Upper Paran River (Brazil): Geomorphology, sedimentology and paleoclimatology, *Quat. Int.*, *21*, 143-161, 1994.

Stevaux, J. C., and M. L. Dos Santos, Paleohydrological changes in the Upper Paraná River: Brazil, during the Late Quaternary: A facies approach, in *Palaeohydrology and Environmental Change*, edited by G. Benito, V. R. Baker and K. J. Gregory, pp. 273-258, Wiley, New York, 1998.

Stocker, T. F., and A. Schmittner, Influence of CO_2 emission rates on the stability of the thermohaline circulation, *Nature, 388*, 862-865, 1997.

Stoner, J. S., J. E. T. Channell, C. Hillaire-Marcel, and C. Kissel, Geomagnetic paleointensity and environmental record from Labrador Sea core MD95-2024:Global marine sediment and ice core chronostratigraphy for the last 110 kyr, *Earth Planet. Sci. Lett., 183*, 161-177, 2000.

Stott, L. D., T. Bunn, M. Prokopenko, C. Mahn, J. Gieskes, and J. M. Bernhard, Does the oxidation of methane leave an isotopic fingerprint in the geologic record?, *Geochem. Geophys. Geosyst., 3*, 10.1029/2001GC000196, 2002.

Street, F. A., and A. T. Grove, Global maps of lake-level fluctuations since 30,000 yr B.P., *Quat. Res., 12*, 83-118, 1979.

Street-Perrott, F. A., Tropical wetland sources, *Nature, 355*, 23-24, 1992.

Stuiver, M., and P. M. Grootes, GISP2 oxygen isotope ratios, *Quat. Res., 53*, 277-284, 2000.

Stuiver, M., P. M. Grootes, and T. F. Braziunas, The GISP2 delta ^{18}O climate record of the past 16,500 years and the role of the sun, ocean, and volcanoes, *Quat. Res., 44*, 341-354, 1995.

Stute, M., M. Forster, H. Frischkorn, A. Serejo, J. F. Clark, P. Schlosser, W. S. Broecker, and G. Bonani, Cooling of tropical Brazil (5°C) during the last glacial maximum, *Science, 269*, 379-383, 1995.

Suess, E., M. E. Torres, G. Bohrmann, R. W. Collier, J. Greinert, P. Linke, G. Rehder, A. Trehu, K. Wallmann, G. Winckler, and E. Zuleger, Gas hydrate destabilization: Enhanced dewatering, benthic material turnover and large methane plumes at the Cascadia convergent margin, *Earth Planet. Sci. Lett., 170*, 1-15, 1999.

Suess, E., M. E. Torres, G. Bohrmann, R. W. Collier, D. Rickert, C. Goldfinger, P. Linke, A. Heuser, H. Sahling, K. Heeschen, C. Jung, K. Nakamura, J. Greinert, O. Pfannkuche, A. Trehu, G. Klinkhammer, M. J. Whiticar, A. Eisenhauer, B. Teichert, and M. Elvert, Sea floor methane hydrates at Hydrate Ridge, Cascadia margin, in *Natural Gas Hydrates: Occurrence, Distribution, and Detection*, edited by C. K. Paull and W. P. Dillon, *Geophys. Monogr. Ser., 124*, pp. 87-98, AGU, Washington D.C., 2001.

Summerhayes, C. P., B. D. Bornhold, and R. W. Embley, Surficial slides and slumps on the continental slope and rise of south west Africa: A reconnaissance study, *Mar. Geol., 31*, 265-277, 1979.

Sun, X., and X. Li, A pollen record of the last 37 ka in deep-sea core 17940 from the northern slope of the South China Sea, *Mar. Geol., 156*, 227-244, 1999.

Sun, X., and Y. Luo, Pollen record of the last 28 ka from deep sea sediments of the northern South China Sea, *Sci. China, Ser. D, 44*, 879-888, 2001.

Swart, P. K., U. G. Wortmann, R. M. Mitterer, M. J. Malone, P. L. Smart, D. A. Feary, and A. C. Hine, Hydrogen sulfide-rich hydrates and saline fluids in the continental margin of South Australia, *Geology, 28*, 1039-1042, 2000.

Tada, R., T. Irino, and I. Koizumi, Land-ocean linkages over orbital and millennial timescales recorded in the late Quaternary sediments of the Japan Sea, *Paleoceanography, 14*, 236-247, 1999.

Takahashi, K., The Bering and Okhotsk Seas: Modern and past paleoceanographic changes and gateway impact, *J. Asian Earth Sci.*, *16*, 49-58, 1998.

Talley, L. D., An Okhotsk Sea water anomaly: Implications for ventilation in the North Pacific, *Deep Sea Res., Part A*, *38*, 171-190, 1991.

Talley, L. D., Distribution and formation of North Pacific Intermediate Water, *J. Phys. Oceanogr.*, *23*, 517-537, 1993.

Talley, L. D., Some aspects of ocean heat transport by the shallow, intermediate and deep overturning circulations, in *Mechanisms of Global Climate Change at Millennial Time Scales*, edited by P. U. Clark, R. S. Webb and L. D. Keigwin, *Geophys. Monogr. Ser.*, *112*, pp. 1-22, AGU, Washington D.C., 1999.

Tarnocai, C., The amount of organic carbon in various soil orders and ecological provinces of Canada, in *Soil Processes and the Carbon Cycle*, edited by R. Lal, J. M. Kimble, R. F. Follett and B. A. Stewart, pp. 81-101, CRC Press, Boca Raton, 1998.

Tathy, J. P., R. A. Delmas, A. Marenco, B. Cros, M. Labat, and J. Servant, Methane emission from flooded forest in Central Africa, *J. Geophys. Res.*, *97*, 6159-6168, 1992.

Taylor, J. A., The peatlands of Great Britain and Ireland, in *Mires: Swamp, Bog, Fen, and Moor: Regional Studies*, edited by A. J. P. Gore and D. W. Goodall, *Ecosyst. World*, *4B*, pp. 1-46, Elsevier, New York, 1983.

Taylor, K., Rapid climate change, *Amer. Sci.*, *87*, 320-328, 1999.

Taylor, K. C., G. W. Lamorey, G. A. Doyle, R. B. Alley, P. M. Grootes, P. A. Mayewski, J. W. C. White, and L. K. Barlow, The 'flickering switch' of late Pleistocene climate change, *Nature*, *361*, 432-436, 1993.

Teller, J. T., J. Risberg, G. Matile, and S. Zoltai, Postglacial history and paleoecology of Wampum, Manitoba, a former lagoon in the Lake Agassiz basin, *Geol. Soc. Am. Bull.*, *112*, 943-958, 2000.

Thevenon, F., A. Vincens, D. Williamson, E. Bard, L. Beauford, P. Dessouillez, G. Buchet, and Y. Ternois, 4000 years of fire activity and vegetation changes from Lake Massoleo (Tanzania), paper presented at International Union for Quaternary Research XV International Congress, Durban, South Africa, 1999.

Thompson, A. M., J. A. Chappellaz, I. Y. Fung, and T. L. Kucsera, The atmospheric CH_4 increase since the Last Glacial Maximum (2). Interactions with oxidants, *Tellus*, *45B*, 242-257, 1993.

Thompson, K., and A. C. Hamilton, Peatlands and swamps of the African continent, in *Mires: Swamp, Bog, Fen and Moor: Regional Studies*, edited by A. J. P. Gore and D. W. Goodall, *Ecosyst. World*, *4B*, pp. 331-373, Elsevier, New York, 1983.

Thompson, L. G., E. Mosely-Thompson, M. E. Davis, P.-N. Lin, K. A. Henderson, J. Cole-Dai, J. F. Bolzan, and K. B. Liu, Late glacial stage and Holocene tropical ice core records from Huascarán, Peru, *Science*, *269*, 46-50, 1995.

Thompson, L. G., M. E. Davis, E. Mosley-Thompson, T. A. Sowers, K. A. Henderson, V. S. Zagorodnov, P.-N. Lin, V. N. Mikhalenko, R. K. Campen, J. F. Bolzan, J. Cole-Dai, and B. Francou, A 25,000-year tropical climate history from Bolivian ice cores, *Science*, *282*, 1858-1864, 1998.

Thorpe, R. B., E. G. Nisbet, K. S. Law, S. Bekki, and J. A. Pyle, Is methane-driven deglaciation consistent with the ice core record?, *J. Geophys. Res.*, *101*, 28627-28635, 1996.

Thorpe, R. B., J. A. Pyle, and E. G. Nisbet, What does the ice-core record imply concerning the maximum climatic impact of possible gas hydrate release at Termination 1A?, in *Gas Hydrates: Relevance to World Margin Stability and Climate Change*, edited by J.-P. Henriet and J. Mienert, *Geol. Soc. Spec. Publ.*, *137*, pp. 319-326, 1998.

Thunell, R. C., and P. G. Mortyn, Glacial climate instability in the Northeast Pacific Ocean, *Nature*, *376*, 504-506, 1995.

Tjia, H. D., The Sunda Shelf, Southeast Asia, *Zeit. Geomorphol.*, *24*, 405-427, 1980.

Toggweiler, J. R., K. Dixon, and W. S. Broecker, The Peru upwelling and the ventilation of the South Pacific thermocline, *J. Geophys. Res.*, *96*, 20467-20497, 1991.

Tréhu, A. M., M. E. Torres, G. F. Moore, E. Suess, and G. Bohrmann, Temporal and spatial evolution of a gas hydrate-bearing accretionary ridge on the Oregon continental margin, *Geology*, *27*, 939-942, 1999.

Tricart, J., El Pantanal: Un ejemplo del impacto de la geomorfologia sobre el ambiente, *Geografia*, *7*, 37-50, 1982.

Tucholke, B., G. Bryan, and E. J., Gas hydrate horizons detected in seismic profiler data from western North Atlantic, *AAPG Bull.*, *61*, 698-707, 1977.

Turcq, B., M. M. N. Pressinotti, and L. Martin, Paleohydrology and paleoclimate of the past 33,000 years at the Tamanduá River, Central Brazil, *Quat. Res.*, *47*, 284-294, 1997.

Turcq, B., R. C. C. Cordiero, A. Sifeddine, F. F. L. Simões Filho, A. O. Da Silva, and J. Capitaneo, Carbon Storage in Amazonia during glacial times (abstract), *EUG Confer. Abstr.*, *10*, 4, 187, 1999a.

Turcq, B., R. C. Cordeiro, A. Sifeddine, F. F. L. Simões Filho, J. J. Abrao, F. Lagoeiro, A. O. da Silva, J. F. Capitaneo, and A. Lima, Was Amazonia a carbon source in the past?, paper presented at International Union for Quaternary Research XV International Congress, Durban, South Africa, 1999b.

Twichell, D. C., D. W. Folger, and H. J. Knebel, Delaware River: Evidence for its former extension to Wilmington Submarine Canyon, *Science*, *195*, 483-485, 1977.

Twitchett, R. J., C. V. Looy, R. Morante, H. Visscher, and P. B. Wignall, Rapid and synchronous collapse of marine and terrestrial ecosystems during the end-Permian biotic crisis, *Geology*, *29*, 351-354, 2001.

Uchupi, E., S. A. Swift, and D. A. Ross, Gas venting and late Quaternary sedimentation in the Persian (Arabian) Gulf, *Mar. Geol.*, *129*, 237-269, 1996.

Ujiié, Y., Mud diapirs observed in two piston cores from the landward slope of the northern Ryukyu Trench, northwestern Pacific Ocean, *Mar. Geol.*, *163*, 149-167, 2000.

Vail, P. R., F. Audemard, S. A. Bowman, P. N. Eisner, and G. Perez-Cruz, The stratigraphic signatures of tectonics, eustasy, and sedimentation—An overview, in *Cyclic Stratigraphy II*, edited by G. Einsele, W. Ricken and A. Seilacher, pp. 617-659, Springer-Verlag, New York, 1991.

Valentine, D. L., Thermodynamic ecology of hydrogen-based syntrophy, in *Symbiosis*, edited by J. Seckbach, Kluwer Academic Pub., Dordrecht, in press.

Valentine, D. L., and W. S. Reeburgh, New perspectives on anaerobic methane oxidation, *Environ. Micrbiol.*, *2*, 477-484, 2000.

Valentine, D. W., E. A. Holland, and D. S. Schimel, Ecosystem and physiological controls over methane production in northern wetlands, *J. Geophys. Res.*, *99*, 1563-1571, 1994.

Valentine, D. L., D. C. Blanton, W. S. Reeburgh, and M. Kastner, Water column methane oxidation adjacent to an area of active hydrate dissociation, Eel River Basin, *Geochim. Cosomochim. Acta*, *65*, 2633-2640, 2001.

Valyashko, G. M., and L. L. Demina, An underwater gas source in the Sea of Okhotsk west of Paramushir Island, *Oceanology*, *27*, 598-602, 1987.

Van den Pol-van Dasselaar, A., and O. Oenema, Methane production and carbon mineralisation of size and density fractions of peat soils, *Soil Biol. Biochem.*, *31*, 877-886, 1999.

Van der Zwaan, G. J., I. A. P. Duijnstee, M. den Dulk, S. R. Ernst, N. T. Jannink, and T. J. Kouwenhoven, Benthic foraminifers: Proxies or problems? A review of paleocological concepts, *Earth Sci. Rev.*, *46*, 213-236, 1999.

van Geen, A., R. G. Fairbanks, P. Dartnell, M. McGann, J. V. Gardner, and M. Kashgarian, Ventilation changes in the Pacific during the last deglaciation, *Paleoceanography*, *11*, 519-528, 1996.

van Kreveld, S., M. Sarnthein, H. Erlenkeuser, P. Grootes, S. Jung, M. J. Nadeau, U. Pflaumann, and A. Voelker, Potential links between surging ice sheets, circulation changes, and the Dansgaard-Oeschger cycles in the Irminger Sea, 60-18 kyr, *Paleoceanography*, *15*, 425-442, 2000.

Van Scoy, K. A., D. B. Olson, and R. A. Fine, Ventilation of North Pacific intermediate waters: The role of the Alaskan Gyre, *J. Geophys. Res.*, *96*, 16801-16810, 1991.

Velichko, A. A., C. V. Kremenetski, O. K. Borisova, E. M. Zelikson, V. P. Nechaev, and H. Faure, Estimates of methane emission during the last 125,000 years in Northern Eurasia, *Global Planet. Change*, *16-17*, 159-180, 1998.

Venz, K. A., D. A. Hodell, C. Stanton, and D. A. Warnke, A 1.0 Myr record of Glacial north Atlantic intermediate water variability from ODP Site 982 in the northeast Atlantic, *Paleoceanography*, *14*, 42-52, 1999.

Verstappen, H. T., On palaeo climates and landform development in Malesia, in *Modern Quaternary Research in Southeast Asia*, edited by G.-J. Bartstra and W. A. Casperie, pp. 3-35, Balkema, Rotterdam, 1975.

Vidal, L., R. R. Schneider, O. Marchal, T. Bickert, T. F. Stocker, and G. Wefer, Link between the North and South Atlantic during the Heinrich events of the last glacial period, *Clim. Dyn.*, *15*, 909-919, 1999.

Vitt, D. H., L. A. Halsey, I. E. Bauer, and C. Campbell, Spatial and temporal trends in carbon storage of peatlands of continental western Canada through the Holocene, *Can. J. Earth Sci.*, *37*, 683-693, 2000.

Voelker, A. H. L., and Workshop Participants, Global distribution of centennial-scale records for Marine Isotope Stage (MIS) 3: A database, *Quat. Sci. Rev.*, *21*, 1185-1212, 2002.

Vogt, P. R., and W.-Y. Jung, Holocene mass wasting on non-polar upper continental slopes—due to post-Glacial ocean warming and hydrate dissociation?, *Geophys. Res. Lett.*, *29*, 55, 2002.

Voigt, B., B. Gabriel, B. Lassonczyk, and M. M. Ghod, Quaternary events at the Horn of Africa, *Berl. Geowiss. Abh.*, *120*, 679-694, 1990.

Waelbroeck, C., J.-C. Duplessy, E. Michel, L. L., D. Paillard, and J. Duprat, The timing of the last deglaciation in North Atlantic climate records, *Nature*, *412*, 724-727, 2001.

Wang, D., and R. Hesse, Continental slope sedimentation adjacent to an ice margin. II. Glaciomarine depositional facies on the Labrador Slope and glacial cycles, *Mar. Geol.*, *135*, 65-96, 1996.

Wang, L., and T. Oba, Tele-connections between east Asian monsoon and the high-latitude climate: A comparison between the GISP 2 ice core record and the high resolution marine records from the Japan and South China seas, *Quat. Res.*, *37*, 211-219, 1998.

Wang, L., M. Sarnthein, H. Erlenkeuser, J. Grimalt, P. Grootes, S. Heilig, E. Ivanova, M. Kienast, C. Pelejero, and U. Pflaumann, East Asian monsoon climate during the Late Pleistocene: High-resolution sediment records from the South China Sea, *Mar. Geol.*, *156*, 245-284, 1999a.

Wang, L., M. Sarnthein, P. M. Grootes, and H. Erlenkeuser, Millennial reoccurrence of century-scale abrupt events of East Asian monsoon: A possible heat conveyor for the global deglaciation, *Paleoceanography*, *14*, 725-731, 1999b.

Wang, X.-C., R. F. Chen, J. Whelan, and L. Eglington, Contribution of "old" carbon from natural marine hydrocarbon seeps to sedimentary and dissolved organic carbon pools in the Gulf of Mexico, *Geophys. Res. Lett.*, *28*, 3313-3316, 2001.

Wang, Y. J., H. Cheng, R. L. Edwards, Z. S. An, J. Y. Wu, C.-C. Shen, and J. A. Dorale, A high-resolution absolute-dated late Pleistocene monsoon record from Hulu Cave, China, *Science*, *294*, 2345-2348, 2001.

Ward, P. D., J. W. Haggart, E. S. Carter, D. Wilbur, H. W. Tipper, and T. Evans, Sudden productivity collapse associated with the Triassic-Jurassic boundary mass extinction, *Science*, *292*, 1148-1151, 2001.

Warford, A. L., D. R. Kosiur, and P. R. Doose, Methane production in Santa Barbara Basin sediments, *Geomicrobiol. J.*, *1*, 117-137, 1979.

Watanabe, T., and M. Wakatsuchi, Formation of 26.8-26.9 σt water in the Kuril Basin of the Sea of Okhotsk as possible origin of North Pacific Intermediate Water, *J. Geophys. Res.*, *103*, 2849-2865, 1998.

Watson, R. T., H. Rodhe, H. Oeschger, and U. Siegenthaler, Greenhouse gases and aerosols, in *Climate Change: The IPCC Scientific Assessment*, edited by J. T. Houghton, G. J. Jenkins and J. J. Ephraums, pp. 41-68, Cambridge University Press, New York, 1990.

Watts, W. A., and B. C. S. Hansen, Environment of Florida in the Late Wisconsin and Holocene, in *Wet Site Archeology*, edited by B. A. Purdy, pp. 307-323, Telford Press, Caldwell, 1988.

Weaver, A. J., C. M. Bitz, A. F. Fanning, and M. M. Holland, Thermohaline circulation: High-latitude phenomena and the difference between the Pacific and Atlantic, *Annu. Rev. Earth Planet. Sci.*, *27*, 231-285, 1999.

Weaver, P. P. E., R. G. Rothwell, J. Ebbing, D. E. Gunn, and P. M. Hunter, Correlation, frequency of emplacement and source directions of megaturbidites on the Madeira Abyssal Plain, *Mar. Geol.*, *109*, 1-20, 1992.

Weber, M. E., M. H. Wiedicke, H. R. Kudrass, C. Hübscher, and H. Erlenkeuser, Active growth of the Bengal Fan during sea-level rise and highstand, *Geology*, *25*, 315-318, 1997.

Wefer, G., P.-M. Heinze, and W. H. Berger, Clues to ancient methane release, *Nature*, *369*, 282, 1994.

Whalen, S. C., W. S. Reeburgh, and C. E. Reimers, Control of tundra methane emission by microbial oxidation, *Ecolog. Studies*, *120*, 257-274, 1996.

White, J. W. C., and E. Steig, Timing of major climate shifts between the poles: Implications for mechanisms of internally forced climate changes, paper presented at PEP 1, Pole-Equator-Pole, Paleoclimate of the Americas, Merida, Venezuela, 16-20 March, 1998.

Whitlock, C., P. Bartlein, V. Markgraf, and A. Ashworth, The midlatitudes of North and South America during the Last Glacial Maximum and early Holocene: Similar paleoclimatic sequences despite differing large-scale controls, in *Interhemispheric Climate Linkages*, edited by V. Markgraf, *391-416*, pp., Academic Press, San Diego, 2001.

Williams, H. E., Notas Geologicas e Economicas sobre o vale do Rio Sao Francisco, *Boletim Servico Geol. Mineral*, *12*, 1-56, 1925.

Williams, M. A. J., D. Adamson, B. Cock, and R. McEvedy, Late Quaternary environments in the White Nile region, Sudan, *Global Planet. Change*, *26*, 305-316, 2000.

Williams, R. T., and R. L. Crawford, Methane production in Minnesota peatlands, *Appl. Environ. Microbiol.*, *4*, 1266-1271, 1984.

Wilson, E. O., *Consilience: The Unity of Knowledge*, 332 pp., Random House, New York, 1998.

Winkler, M. G., P. R. Sanford, and S. W. Kaplan, Hydrology, vegetation, and climate change in the southern Everglades during the Holocene, *Bull. Amer. Paleontol.*, *361*, 57-99, 2001.

Wong, A. P. S., N. L. Bindoff, and J. A. Church, Large-scale freshening of intermediate waters in the Pacific and Indian oceans, *Nature*, *400*, 440-443, 1999.

Woodside, J. M., M. K. Ivanov, and A. F. Limonov, Shallow gas and gas hydrates in the Anaximander Mountains region, eastern Mediterranean Sea, in *Gas Hydrates: Relevance to World Margin Stability and Climate Change*, edited by J.-P. Henriet and J. Mienert, *Geol. Soc. Spec. Publ.*, *137*, pp. 177-193, 1998.

Wright, H. E., Jr., and P. H. Glaser, Postglacial peatlands of the Lake Agassiz Plain, northern Minnesota, in *Glacial Lake Agassiz*, edited by J. T. Teller and L. Clayton, *Geol. Ass. Can. Spec. Paper*, *26*, pp. 375-389, 1983.

Wunsch, C., On sharp spectral lines in the climate record and the millennial peak, *Paleoceanography*, *15*, 417-424, 2000.

Xiao, J. L., Z. S. An, T. S. Liu, Y. Inouchi, H. Kumai, S. Yoshikawa, and Y. Kondo, East Asian monsoon variation during the last 130,000 years: Evidence from the Loess Plateau of central China and Lake Biwa of Japan, *Quat. Sci. Rev.*, *18*, 147-157, 1999.

Xu, W., R. P. Lowell, and E. T. Peltzer, Effect of seafloor temperature and pressure variations on methane flux from a gas hydrate layer: Comparison between current and late Paleocene climate conditions, *J. Geophys. Res.*, *106*, 26413-26423, 2001.

Xu, X. Y., and C. Ruppel, Predicting the occurrence, distribution, and evolution of methane gas hydrate in porous marine sediments, *J. Geophys. Res.*, *104*, 5081-5095, 1999.

Yasuda, I., The origin of North Pacific Intermediate Water, *J. Geophys. Res.*, *102*, 893-909, 1997.

Yasuda, I., K. Okuda, and Y. Shimizu, Distribution and modification of North Pacific Intermediate Water in the Kuroshio-Oyashio interfrontal zone, *J. Phys. Oceanogr.*, *26*, 448-465, 1996.

Yavitt, J. B., G. E. Lang, and D. M. Downey, Potential methane production and methane oxidation rates in peatland ecosystems of the Appalachian Mountains, United States, *Global Biogeochem. Cycles*, *2*, 253-268, 1988.

Yokoyama, Y., T. M. Esat, and K. Lambeck, Coupled climate and sea-level changes deduced from Huon Peninsula coral terraces of the last ice age, *Earth Planet. Sci. Lett.*, *193*, 579-587, 2001.

Yun, J. W., D. L. Orange, and M. E. Field, Subsurface gas offshore of northern California and its link to submarine geomorphology, *Mar. Geol.*, *154*, 357-368, 1999.

Zachos, J. C., and G. R. Dickens, An assessment of the biogeochemical feedback response to the climatic and chemical perturbations of the LPTM, *GFF*, *122*, 188-189, 2000.

Zachos, J., M. Pagani, L. Sloan, E. Thomas, and K. Billups, Trends, rhythms, and aberrations in global climate 65 Ma to present, *Science*, *292*, 686-693, 2001.

Zengler, K., H. H. Richnow, R. Rosselló-Mora, W. Michaelis, and F. Widdel, Methane formation from long-chain alkanes by anaerobic microorganisms, *Nature*, *401*, 266-269, 1999.

Zhang, C. L., Y. Li, J. D. Wall, L. Larsen, R. Sassen, Y. Huang, Y. Wang, A. Peacock, D. C. White, J. Horita, and D. R. Cole, Lipid and carbon isotopic evidence of methane-oxidizing and sulfate-reducing bacteria in association with gas hydrates from the Gulf of Mexico, *Geology*, *30*, 239-242, 2002.

Zheng, Y., A. van Geen, R. F. Anderson, J. V. Gardner, and W. E. Dean, Intensification of the northeast Pacific oxygen minimum zone during the Bølling-Ållerød warm period, *Paleoceanography*, *15*, 528-536, 2000.

Zimov, S. A., Y. V. Voropaev, I. P. Semiletov, S. P. Davidov, S. F. Prosiannikov, F. S. Chapin, M. C. Chapin, S. Trumbore, and S. Tyler, North Siberian lakes: A methane source fueled by Pleistocene carbon, *Science*, *277*, 800-802, 1997.

Zoltai, S. C., and C. Tarnocai, Perennially frozen peatlands in the western Arctic and Subarctic of Canada, *Can. J. Earth Sci.*, *12*, 28-43, 1975.

Zoltai, S. C., and F. C. Pollett, Wetlands in Canada: Their classification, distribution, and use, in *Mires: Swamp, Bog, Fen, and Moor: Regional Studies*, edited by A. J. P. Gore and D. W. Goodall, *Ecosyst. World*, *4B*, pp. 245-268, Elsevier, New York, 1983.

Zoltai, S. C., and D. H. Vitt, Holocene climatic change and the distribution of the peatlands in Western Interior Canada, *Quat. Res.*, *33*, 231-240, 1990.

Zonenshain, L. P., I. O. Murdmaa, B. V. Baranov, A. P. Kuznetsov, V. S. Kuzin, M. I. Kuzmin, G. P. Avdeiko, P. A. Stunzhas, V. N. Lukashin, M. S. Barash, G. M. Valyashko, and L. L. Dyomina, An underwater gas source in the Sea of Okhotsk west of Paramushir Island, *Oceanology*, *27*, 598-602, 1987.

List of Abbreviations and Symbols

AAIW = Antarctic Intermediate Water
B-Å = Bølling-Ållerod
B.P. = Before Present
BSR = bottom-simulating reflector
D/O = Dansgaard/Oeschger
Eg = Exagrams (10^{18} g)
GISP2 = Greenland Ice Sheet Project 2
GRIP = Greenland Ice Core Project
ha = hectare (10^4 m^2)
ka = kilannum (thousand years before present)
kyr = thousand years
LGM = Last Glacial Maximum
Ma = megannum (million years before present)
MAR = mass accumulation rate
MIS = Marine Isotope Stage
myr = million years
NADW = North Atlantic Deep Water
NPIW = North Pacific Intermediate Water
ODP = Ocean Drilling Program
OMZ = oxygen-minimum zone
PDB = Peedee Belemnite (isotopic carbonate standard)
Pg = Petagrams (10^{15} g)
ppbv = parts per billion by volume
ppmv = parts per million by volume
SMOW = Standard Mean Ocean Water (isotopic water standard)
SST = sea-surface temperature
Sv = Sverdrups (10^6 m^{-3} sec^{-1})
Tg = Teragrams (10^{12} g)
‰ = per mil (parts per thousand)
$\delta^{13}C$ = normalized ratio of ^{13}C to ^{12}C relative to a standard in parts per thousand
δD = normalized ratio of ^2H to H relative to a standard in parts per thousand
$\delta^{18}O$ = normalized ratio of ^{18}O to ^{16}O relative to a standard in parts per thousand

Glossary of Terms

Albedo. Reflectivity of a surface; the ratio of the amount of electromagnetic energy reflected by a surface to the amount of energy incident upon it.

Alkenone-derived SST (sea-surface temperatures). Sea-surface temperatures determined using the degree of unsaturation of fossil long-chain alkenones from certain marine algae such as coccolithophorids. Long-chain polyunsaturated ketones (alkenones) are sensitive to temperature changes during biological growth, particularly the C_{37} ketone (containing 37 carbon atoms). A linear relationship exists between the abundance of C_{37} ketones (the $U^{k'}_{37}$ index) and SST.

Archaea. An important and ancient group of prokaryotic microbes, distinguished from bacteria by lipid cell membranes composed of isoprenes instead of fatty acids. This group includes methanogens, methanotrophs, extreme halophiles, extreme thermophiles, and the sulfate-reducing bacteria.

Authigenic. Sediments and/or minerals that precipitate in place.

Benthic. Ocean or lake floor environments and associated bottom-dwelling life forms.

Biomarkers. Distinctive organic compounds, usually lipids, used to identify the presence of specific microbes or algal species (see hopanoid diplopterol).

Bond Cycles. Distinct ~7.4 kyr oscillation with each cycle consisting of several stadial-interstadial episodes of decreasing duration or amplitude. Each cycle concludes with a severely cold Heinrich Event marked by an interval of North Atlantic ice rafting followed immediately by an abrupt warming into the next cycle.

Bottom-simulating reflector (BSR). Distinct seismic discontinuity described in marine and lake sediment sequences that parallels ocean topography. This, most frequently, represents the phase boundary marking the top of the free gas zone beneath a solid methane hydrate-rich layer.

Cenozoic. Geologic era extending from the end of the Cretaceous Period to the present, representing the last 65 million years.

Clathrate. Nonstoichiometric compound with a lattice of water molecules with gas molecules (e.g., CH_4, CO_2, H_2S) occupying cavities (or cages) within the lattice.

Dansgaard/Oeschger Cycles. Millennial-scale climate oscillations during the last glacial episode between warm interstadials and cold stadials (periodicity of ~1400 to 1500 yr). Named after W. Dansgaard and H. Oeschger who first discovered them in Greenland ice core records.

Deglaciation. Interval of melting of the polar ice sheets following glacial maxima including glacial terminations and early interglacials. The last deglacial interval occurred from ~18 ka to 7 ka with particularly strong ice sheet decay between ~15 and 9 ka.

Denitrification. A process in suboxic sedimentary and aqueous conditions by which bacteria metabolically convert nitrate (NO_3^-) to N_2 gas, where NO_3^- serves as the electron acceptor. This process results in nitrogen fractionation (bacteria preferentially uptake ^{14}N), leaving a record of the process in sedimentary organic matter.

Epibenthic. Benthic organisms that live on the surface of the ocean floor, rather than within the sediments.

Feedback. A process in which factors that produce a result are themselves strengthened or reinforced (positive feedback) or weakened (negative feedback) by responding factors; when an output of a system also serves as an input to cause further change to the system

Foraminifera. Order of sarcodine protistids with a test usually constructed of calcite or of agglutinated particles. A very important microfossil group for paleoceanography and biostratigraphy, especially for the Cenozoic Era.

Gas Hydrates. See Clathrates.

Glacial Terminations. Abrupt, punctuated episodes of major near-global warming and ice sheet melting that terminate glacial episodes. The last glacial episode was terminated by two such abrupt steps, Termination 1A at ~14.7 ka and Termination 1B at ~11.5 ka.

Glacial. One of many latest Pliocene through Quaternary climate intervals that occurred between interglacial epsodes, marked by extensive ice sheet expansions and relatively cold climate over broad areas of the Earth.

Heinrich Events. Relatively brief (100 to 500 yr) intervals marked by massive iceberg discharge from disintegrating ice sheets and associated meltwater episodes in high north Atlantic latitudes. These events are recorded as thick (several meters) sediment layers (Heinrich Layers) immediately prior to abrupt, major warming shifts of the last glacial episode.

Holocene. Geologic epoch of the Quaternary Period representing the last ~11.5 ka of earth history, corresponding to the most recent interglacial episode.

Hopanoid diplopterol. Lipid biomarker, present in aerobic bacteria and especially abundant in methanotrophic bacteria. Used as an indicator of aerobic methanotrophy.

Infaunal. Benthic organisms that live within rather than on bottom sediments.

Insolation. The amount of solar radiation reaching a specific area of the Earth's surface.

Interglacial. One of many latest Pliocene through Quaternary climate episodes that occurred between glacial intervals, marked by relatively warm temperatures over broad areas of the Earth and of sufficient duration to have led to reduced ice extent much like that of the present day.

Intermediate Waters. Cool water masses between the permanent thermocline and Deep Water (~400 to ~1500 m) produced in the Arctic and Antarctic regions.

Interstadials. Warm climate episodes of the last ice age, each lasting hundreds to several thousand years, marked by glacial retreat and separated by cold, stadial episodes.

Intertropical Convergence Zone. Atmospheric division between northern and southern tropics forming a zone mostly north of the equator marked by convergence of surface (trade) winds in a region of lowest atmospheric pressure associated with warmest surface temperatures. Extreme upwelling of warm air in this zone leads to maxima in rainfall and cloudiness of convective origin.

Methane hydrate. The ice-like state of CH_4 and H_2O, formed as CH_4 molecules captured within a cage of water molecules, produced under conditions of low temperatures, high pressure and sufficient gas concentrations. See clathrate.

Methanogenesis. The process of methane formation by microbes under anoxic conditions.

Methanotrophy. The process of methane oxidation by microbes.

Milankovitch Cycles. Astronomical cycles of climate change that resulted from fluctuations in the seasonal and geographic distribution of insolation caused by variations in the Earth's orbital elements: eccentricity, obliquity or tilt of the rotational axis, and longitude of the perihelion (precession). Named after Yugoslav mathematician M. Milankovitch (1870-1958).

Neoproterozoic. The youngest subdivision of the Precambrian, immediately preceding the Cambrian and representing the interval between ~750 Ma and 548 Ma.

Nepheloid Layer. A layer, hundreds of meters thick within the ocean's water column, containing concentrations of fine, suspended sediment, the depth of which is controlled by density stratification.

Orbital-scale Cycles. See Milankovitch cycles.

Oxygen-minimum zone. Level at upper intermediate depths (~200 to 1000 m) in the ocean marked by lowest dissolved oxygen, high nutrients and associated with organic-rich sediments where it intercepts the sea floor.

Paleocene. First geologic epoch of the Cenozoic, following the Cretaceous Period and before the Eocene Epoch (~65 to 54 Ma).

Paleocryological. Changes on geological time scales in the cryosphere (ice, snow, permafrost).

Palustrine. Pertaining to deposition in a marsh-like environment.

Peat. Deposit of partially decomposed organic matter (largely plants) in a largely anoxic, water-saturated environment such as a bog.

Permafrost. Continental near-surface deposits that have been at temperatures below freezing for extended periods of time; perennially frozen ground.

Planktonic. Pelagic organisms that float, drift or swim weakly and of the environment in which they live.

Pluvial. Climate or episode or process resulting from high rainfall.

Pockmarks. Crater-like depressions on the ocean floor commonly found on continental margins. Their diameter varies greatly from meters to 100 meters with depths up to 10 m.

Productivity. The net rate of production of organic (biological) matter from inorganic sources of carbon in oceans or lakes, chiefly by photosynthetic plankton. Typically expressed in (mass of carbon assimilated)(area)$^{-1}$(time)$^{-1}$.

Quaternary. Interval of geologic time representing the last ~1.8 million years, consisting of the Pleistocene up to 11.5 ka and the Holocene since that time. The <u>late</u> Quaternary represents the last ~0.8 kyr of this period and exhibits the largest climatic and glacial oscillations of the classic ice age.

Speleothems. Calcium carbonate mineral deposits formed by precipitation from water in caves (e.g., stalactities, stalagmites and crusts).

Stadials. Cold climate episodes of the last ice age, each lasting hundreds to several thousand years and marked by glacial advance.

Stratosphere. An outer layer of the atmosphere above the troposphere between ~10 – 50 km.

Thermohaline Circulation. Motion of water in the deep ocean caused by density differences due to variations in temperatures and salinity and different from surface circulation which is wind driven.

Troposphere. The lower part of the atmosphere (up to 10 – 16 km) marked by rapid upward decrease in temperature, cloud formation and active convection.

Turbidite. A typically graded, sediment sequence largely deposited from a turbidity current, which is a short-lived, gravity-driven bottom current consisting of a mixture of suspended sediment and water of density greater than that of the surrounding water. Suspension of sediment is maintained by internal turbulence. Turbidity currents are of major importance in the transportation of terrigenous sediments from shallow water to the deep ocean basins.

Upwelling. The rising of cold subsurface waters toward the ocean surface. Since subsurface waters are often nutrient-rich, upwelling can lead to increased biological productivity at or near the ocean surface.